THE STEPHEN BECHTEL FUND

IMPRINT IN ECOLOGY AND THE ENVIRONMENT

The Stephen Bechtel Fund has
established this imprint to promote
understanding and conservation of
our natural environment.

The publishers gratefully acknowledge the generous contribution to this book provided by the Stephen Bechtel Fund.

COMPARING FUTURES FOR THE SACRAMENTO–SAN JOAQUIN DELTA

COMPARING FUTURES FOR THE SACRAMENTO–SAN JOAQUIN DELTA

Jay R. Lund, Ellen Hanak, William E. Fleenor, William A. Bennett, Richard E. Howitt, Jeffrey F. Mount, and Peter B. Moyle

UNIVERSITY OF CALIFORNIA PRESS Berkeley Los Angeles London
PUBLIC POLICY INSTITUTE OF CALIFORNIA

University of California Press, one of the most distinguished university presses in the United States, enriches lives around the world by advancing scholarship in the humanities, social sciences, and natural sciences. Its activities are supported by the UC Press Foundation and by philanthropic contributions from individuals and institutions. For more information, visit www.ucpress.edu.

Freshwater Ecology Series, Volume 3
For digital edition, see www.ucpress.edu.

University of California Press
Berkeley and Los Angeles, California

University of California Press, Ltd.
London, England

Library of Congress Cataloging-in-Publication Data

Comparing futures for the Sacramento-San Joaquin Delta / Jay R. Lund . . . [et al.].
p. cm.—(Freshwater ecology series ; vol. 3.)
Includes bibliographical references and index.
ISBN 978-0-520-26197-6 (cloth : alk. paper)—
ISBN 978-0-520-94537-1 (electronic)
1. Water quality management—California—Delta Region. 2. Estuarine ecology—California—Delta Region. 3. Water-supply—California—Delta Region—Forecasting. 4. Water diversion—Environmental aspects—California—Delta Region—Forecasting. 5. Delta Region (Calif.)—Environmental conditions. 6. Water-supply—California—Delta Region—Management. 7. Environmental management—California—Delta Region. I. Lund, Jay R.

TD225.D29C66 2010
333.91'64-dc22 2009042977

16 15 14 13 12 11 10
10 9 8 7 6 5 4 3 2 1

The paper used in this publication meets the minimum requirements of ANSI/NISO Z39.48-1992 (R 1997)(*Permanence of Paper*).

Cover image: A landscape of transitions—deeply subsided Delta islands with remnants of original marshland, common characteristics of developed deltas worldwide. Photo by the State of California Department of Water Resources.

CONTENTS

CONTRIBUTORS

William A. Bennett
Center for Watershed Sciences
University of California, Davis
wabennet@ucdavis.edu

William E. Fleenor
Department of Civil and Environmental Engineering
University of California, Davis
wefleenor@ucdavis.edu

Ellen Hanak
Public Policy Institute of California
San Francisco, California
hanak@ppic.org

Richard E. Howitt
Department of Agricultural and Resource Economics
University of California, Davis
howitt@primal.ucdavis.edu

Jay R. Lund
Department of Civil and Environmental Engineering
University of California, Davis
jrlund@ucdavis.edu

Jeffrey F. Mount
Department of Geology
University of California, Davis
mount@geology.ucdavis.edu

Peter B. Moyle
Department of Wildlife, Fish, and Conservation Biology
University of California, Davis
pbmoyle@ucdavis.edu

PREFACE

> One gains nothing . . . by starting out with the question, "What is acceptable?" And in the process of answering it, one gives away the important things, as a rule, and loses any chance to come up with an effective, let alone with the right, answer.
>
> PETER F. DRUCKER (1967), *The Effective Executive*

In the American West, and much of the world, the golden era of water development is over. No longer can dams, diversions, canals, levees, dikes, and ditches be built without regard for the environment. Today's landscapes are saturated with water infrastructure and human land uses. Often, costs of this infrastructure have escalated beyond their benefits. In particular, natural ecosystems have been reduced to small remnants of their historical extent. As these ecosystems have declined, their value has increased, to a point where conflicts among diverse, vested interest groups have become highly visible and often seemingly irresolvable.

As the costs of expanding water infrastructure increase dramatically, additional water supply and flood reduction benefits from new projects are often modest. Advances in water use efficiency, groundwater banking, water treatment, floodplain management, and water markets are typically more effective from both economic and environmental perspectives, particularly when integrated with existing infrastructure. In California, a remarkable transition in water management has occurred as these advances in water operations and water use efficiency substitute for major infrastructure expansions. This combination of sound infrastructure and creative management serves both a growing economy and the growing desire to protect the natural environment.

Yet all is not well. Water problems are always changing. In California, demands for better water management continue to grow, diversify, and challenge a water system magnificently conceived almost a century ago for more limited purposes. The environment continues to deteriorate largely as a result of human activities, as indicated by declining native fish populations. And external forces (climate change and sea-level rise), as well as accumulated effects of long-term problems (land subsidence of peat soils, salinization of irrigated soils in arid regions, nitrate pollution of groundwater, and land use effects on habitats) are pushing aquatic ecosystems into perilous conditions and threatening regional economies.

This book examines an issue central to California's strategic water balancing problem: how to move the Sacramento–San Joaquin Delta to a more sustainable and desirable state. The Delta, together with San Francisco Bay, forms the largest estuary on the west coast of the Americas and is the largest single source in California's water supply system. The Delta faces inevitable changes that make present water policies unsustainable. Rising sea level, continued land subsidence, earthquakes, invasive species, and a worsening climate for floods are among the changes that will overwhelm current Delta management for local agriculture and statewide water supply. With major undesirable consequences foreseen for almost all stakeholder interests, current Delta management implies its own demise.

This book is the result of a true multidisciplinary examination of the Delta over the past three years. The authors include two economists (Ellen Hanak and Richard Howitt), two engineers (William Fleenor and Jay Lund), two biologists (Peter Moyle and William Bennett), and a geologist (Jeffrey Mount), all of whom have a long and diverse history of involvement in California water issues. The book is based on extensive investigations and analysis conducted over a three-year period, including two major reports and ten technical reports published by the Public Policy Institute of California.

We start from the premise that no matter what choices society makes, the Delta of tomorrow will become very different from today. Its static, traditional infrastructure and management are not serving California well. We then address the broad questions: Can a Delta ecosystem be created that sustains desirable species while also maintaining a substantial supply of water for human use? Can California actually manage such a profound transformation? We show that rational, scientifically-supported, and cost effective solutions are available. However, without leadership and long-term thinking, we suspect that the Delta will instead fail into some less desirable environmental and economic condition. With this book we present our technical consensus. We think this work demonstrates that independent research conducted in universities and research institutes can help resolve some of the world's real problems. We hope readers will also get a glimpse of the great intellectual thrill and satisfaction that we, a diverse group of engaged scholars, found in working together on challenging problems.

ACKNOWLEDGMENTS

Much of this work would be weaker without the diligent efforts of our students, research associates, and postdoctoral researchers: Dane Behrens, Wei-Hsiang Chen, Christina Connell, Kevin Fung, Kristine Haunschild, Kaveh Madani, Josue Medellin, Marcelo Olivares, Robyn Suddeth, Sarah Swanbeck, and Stacy Tanaka. They are coauthors of and contributors to many of the technical reports that support this book.

We thank those who contributed by participating in various technical workshops and one-on-one discussions that helped our thinking. These include Kome Ajise, Jamie Anderson, Elaine Archibald, Chuck Armor, George Basye, Gary Bobker, Fabian Bombardelli, Ross Boulanger, Alf Brandt, Jon Burau, John Cain, Tina Cannon Leahy, Pam Carder, Jeff Carroll, Douglas Chun, Francis Chung, Tom Clark, Marci Coglianese, Gilbert Cosio, Joe Countryman, William Craven, Cathy Crother, Martha Davis, John DeGeorge, Susan Dell'Osso, Holly Doremus, John Durand, Allison Dvorak, Chris Enright, Tom Erb, Linda Fiack, Jamie Fordyce, Tony Francois, Dave Fullerton, Gerald Galloway, Greg Gartrell, Alan Gordon, Mark Gowdy, Tom Graff, Dorothy Green, Joe Grindstaff, Les Grober, Marianne Guerin, Sergio Guillen, David Guy, Les Harder, Ann Hayden, Roger Henderson, Bruce Herbold, Alex Hildebrand, Mary Hildebrand, Doug Holland, Paul Hutton, Jerry Johns, Tariq Kadir, Randy Kanouse, Steve Kasower, Stuart Krasner, David Lawson, Clifford Lee, George Lemker, Owen Lu,

Terry Macaulay, Steve MacCauley, Clyde MacDonald, Senator Michael Machado, Steve McCarthy, Rod Meade, Jerry Meral, B. J. Miller, Michael Mirmazaheri, Laura King Moon, Anson Moran, Scott Morgan, Michelle Morrow, David Mraz, Armin Munevar, Nancee Murray, Phil Nails, Barry Nelson, Chris Neudeck, Mary Nichols, Dennis O'Connor, Tim Quinn, Peter Rabbon, Alex Rabidoux, Richard Rachielle, Michael Ramsbotham, Dianne Riddle, Spreck Rosekranz, Said Salah-Mars, Anthony Saracino, Curtis Schmutte, Robert Schroeter, Raymond Seed, Deanna Sereno, K. T. Shum, Pete Smith, Tara Smith, Ted Sommer, Francis Spivy-Weber, Gregory Thomas, Mike Wade, Walt Wadlow, Brent Walthall, Bethany Westfall, Victoria Whitney, Rebecca Willis, Greg Wilson, Leo Winternitz, Wim Kimmerer, Betty Yee, Peter Zhon, Tom Zuckerman, members of the Interagency Ecological Program's Estuarine Ecology Team, and several state and federal levee engineers who asked not to be identified.

We are grateful to many reviewers of earlier, related work for improving substantially our thinking and writing: Ignatius Anwanyu, Elaine Archibald, Elisa Barbour, Alf Brandt, David Briggs, Jon Burau, John DeGeorge, Holly Doremus, Andrew Draper, Chris Enright, Terry Erlewine, David Fullerton, Gerald Galloway, Greg Gartrell, Brian Gray, Les Grober, Kamyar Guivetchi, Maurice Hall, Sam Harader, Steve Hatchett, Jon Haveman, Bruce Herbold, Ray Hoagland, Paul Hutton, Kenneth Kirby, Jed Kolko, Owen Lu, Terry Macaulay, Erin Mahaney, Judith Meyer, B. J. Miller, Laura King Moon, Armin Munévar, Kent Nelson, Fred Nichols, Mary Nichols, Peter Rabbon, Deborah Reed, Kenneth Rose, Spreck Rosekrans, Anthony Saracino, Curtis Schmutte, K. T. Shum, Theodore Sommer, Michael Teitz, Robert Wilkinson, Leo Winternitz, David Zilberman, and one reviewer who wished to remain anonymous.

We also thank Gary Bjork and Lynette Ubois (PPIC), Patricia Bedrosian (RAND), and Lauren Muscatine (UC-Davis) for editorial support and Janice Fong (UC-Davis Geology Department) for designing the maps of the Delta used here.

This work was supported financially by the David and Lucile Packard Foundation, Steven D. Bechtel Jr., and the Public Policy Institute of California. We are extremely grateful for this financial support and for the clerical and office support of the John Muir Institute for the Environment and its Watershed Science Center on the University of California–Davis campus.

We alone are responsible for any remaining errors and for all interpretations of the material presented here. Research publications reflect the views of the authors and do not necessarily reflect the views of the staff, officers, or board of directors of the Public Policy Institute of California or the University of California.

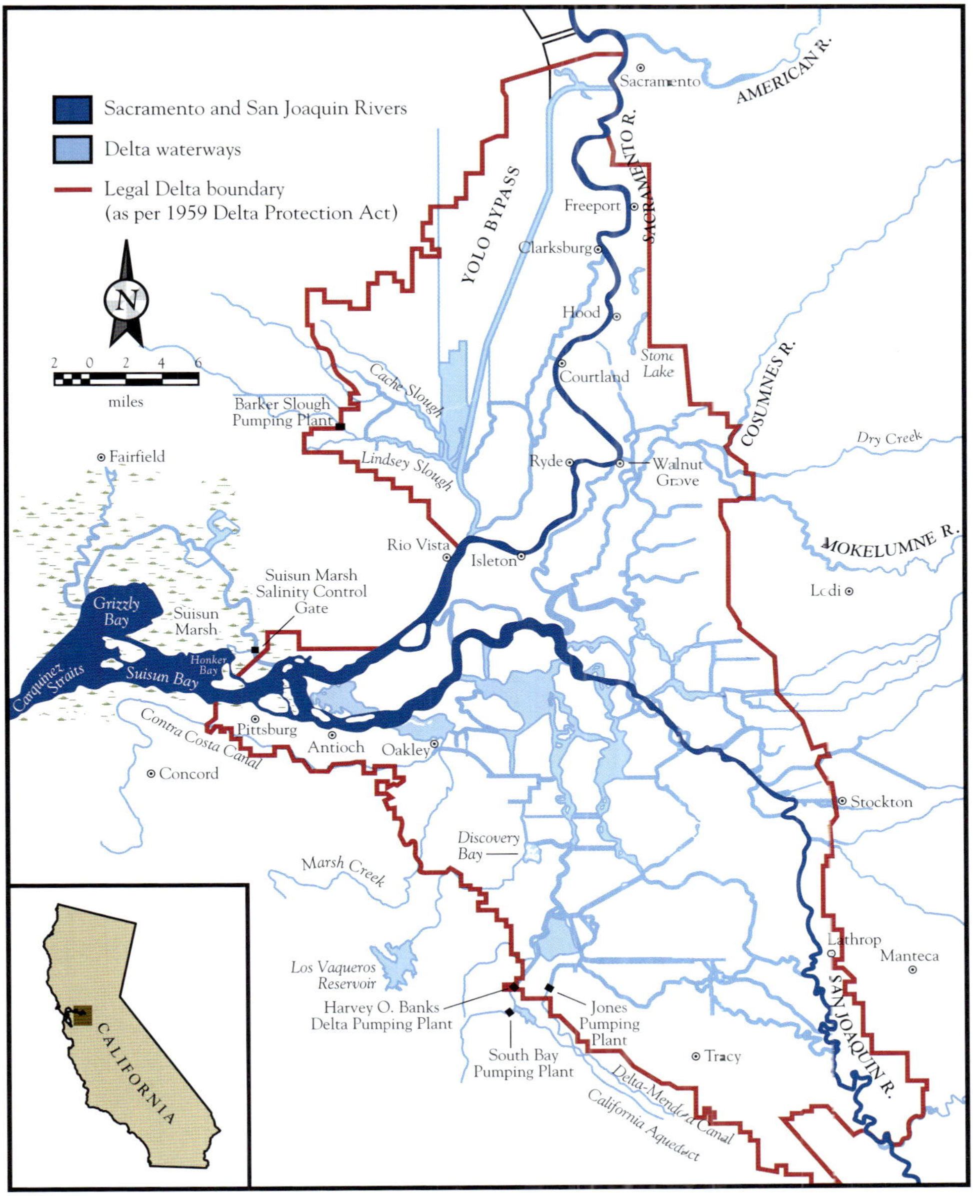

The Sacramento–San Joaquin Delta.

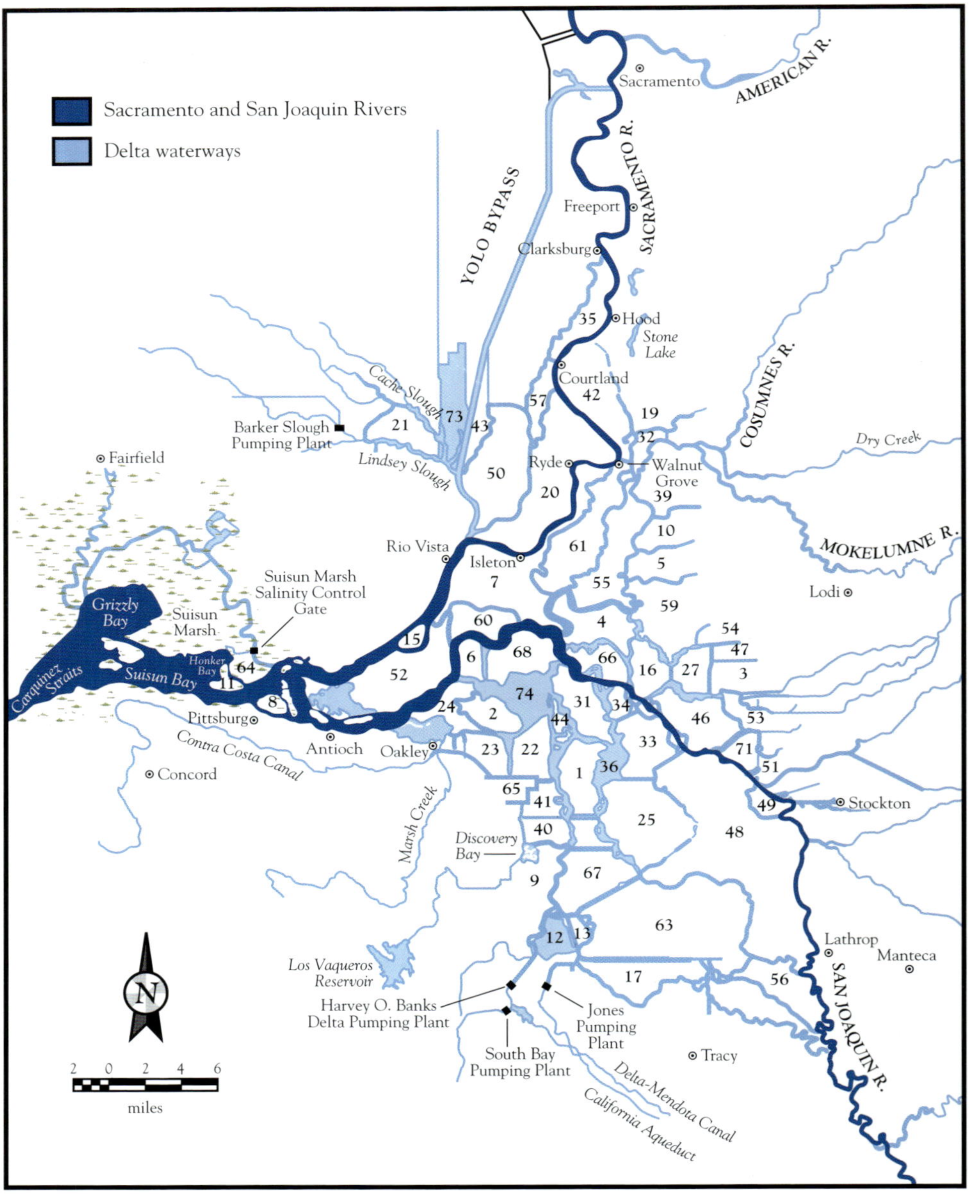

Delta Islands.

DELTA ISLANDS MAP KEY

1. Bacon Island
2. Bethel Tract
3. Bishop Tract
4. Bouldin Island
5. Brack Tract
6. Bradford Island
7. Brannan-Andrus Island
8. Browns Island
9. Byron Tract
10. Canal Ranch
11. Chipps Island
12. Clifton Court Forebay
13. Coney Island
14. Deadhorse Island*
15. Decker Island
16. Empire Tract
17. Fabian Tract
18. Fay Island*
19. Glanville Tract
20. Grand Island
21. Hastings Tract
22. Holland Tract
23. Hotchkiss Tract
24. Jersey Island
25. Jones Tract
26. Kimball Island*
27. King Island
28. Little Franks Tract*
29. Little Mandeville Island*
30. Little Tinsley Island*
31. Mandeville Island
32. McCormack Williamson Tract
33. McDonald Tract
34. Medford Island
35. Merritt Island
36. Mildred Island
37. Netherlands*
38. Neville Island*
39. New Hope Tract
40. Orwood Tract
41. Palm Tract
42. Pierson District
43. Prospect Island
44. Quimby Island
45. Rhode Island*
46. Rindge Tract
47. Rio Blanco Tract
48. Roberts Island
49. Rough and Ready Island
50. Ryer Island
51. Sargent Barnhart Tract
52. Sherman Island
53. Shima Tract
54. Shin Kee Tract
55. Staten Island
56. Stewart Tract
57. Sutter Island
58. Sycamore Island*
59. Terminous Tract
60. Twitchell Island
61. Tyler Island
63. Union Island
64. Van Sickle Island
65. Veale Tract
66. Venice Island
67. Victoria Island
68. Webb Tract
69. Winter Island*
70. Woodward Island
71. Wright-Elmwood Tract
73. Liberty Island
74. Franks Tract

*Numbers not shown on map

I

INTRODUCTION

> It took me eight days before I could find the entrance of the Sacramento, as it is very deceiving and very easy to pass by.
>
> JOHN SUTTER, *The Diary of Johann August Sutter,* 1838–1839 entry

Throughout the world, and particularly in the American West, people are learning how to remanage natural resource and environmental systems, which they had thought of as fully developed and sustainable. In many cases, the old assumptions are proving false. External forces such as sea-level rise, climate change, economic globalization, population growth, and rising concern for the natural environment all impose changes on the management of these systems. In some cases, the internal dynamics, however well intentioned, are also proving unsustainable—with outcomes such as soil erosion, accumulation of pollutants in soils, and groundwater deterioration imposing changes in management over time. This complex confluence of changes in circumstances and expectations should encourage societies periodically to rethink how to manage natural systems. Yet such questioning can be difficult when people have many years of experience and investment in the past.

The Sacramento–San Joaquin Delta is part of the largest estuary on the West Coast of the Americas, providing a home to roughly 50 species of fish and close to 300 species of birds, mammals, and reptiles. The Delta is also the largest single source of California's water supply, channeling water from Northern California's watersheds to two-thirds of the state's households and millions of acres of southern Central Valley farmlands. Locally, the Delta also supports a productive agricultural and recreational economy.

The Delta's ecological, water supply, and local land use functions are in crisis, with crashing populations of native fish species and increasing risks of a catastrophic failure of fragile levees—an event that could severely disrupt the state's water supply as well as local activities. Because the current water supply system has helped change the Delta ecosystem in unfavorable ways, water exports are also susceptible to cutbacks, to protect endangered fish species. This combination of extreme risks—to the state's water supply, the estuarine ecology, and the local economy—make the Delta the foremost water management problem facing California.

Strategies to manage the Delta to satisfy competing interests have been discussed and debated for almost 100 years, at times leading to acrimonious divisions between Northern and Southern California, environmental and economic interests, and agricultural and urban sectors. Recently, the Delta has again taken center stage in debates on California water policy. Research and actual levee failures have exposed the Katrina-level fragility of 1,100 miles of levees, on which both Delta land uses and water supply systems currently depend. In addition, dramatic population declines have occurred among several fish species that depend on the Delta. As awareness of these risks has heightened, it has also exposed weaknesses in the institutional framework for governing the Delta watershed. By late 2004, the state and federal government sponsored a stakeholder-driven process known as CALFED—established a decade earlier to mediate conflict and to "fix" the problems of the Delta—had begun to unravel. As this informal truce among competing interests eroded, lawsuits have filled the gaps left by a lack of consensus on management strategies and options. For the past 70 years, California's official policy has been to maintain the Delta as a freshwater system through a program of water flow regulation, supported by the maintenance of agricultural levees. This approach now appears near or past the end of its useful life, given the deterioration of the Delta's ecosystem and levees, as well as the rising consequences of levee failure.

This book is about finding better solutions to Delta problems. We do not pretend to offer a perfect, comprehensive solution; 100 years of history would argue that kind of solution is unlikely. Indeed, it may be that different Delta strategies are appropriate for different periods in California's development. Instead, our aim is to launch a serious, scientific search and comparison of potential long-term strategies, and provide some broad guidance for the coming decades. This analysis is wide-ranging and integrated, with a focus on the future and best management of the Delta and its landscape and inflows for environmental and economic purposes.

WHAT IS THE DELTA?

The Delta is a web of channels and reclaimed islands at the confluence of the Sacramento and San Joaquin rivers. It forms the eastern portion of the wider San Francisco Estuary, which includes the San Francisco, San Pablo, and Suisun bays, and it collects water from California's largest watershed, which encompasses roughly 45 percent of the state's surface area and stretches from the eastern slopes of the Coast Range to the western slopes of the Sierra Nevada. It resembles other deltas of the world in that it is at the mouth of rivers, receives sediment deposits from these rivers, and was once a vast tidal marsh. The Sacramento–San Joaquin Delta fundamentally differs from other delta systems, however, in that it is not formed primarily by sediment deposits from upstream. Instead, it is a low-lying region where sediment from the watershed commingled with vast quantities of organic matter deposited by tules and other marsh plants. For some 6,000 years, sediment accumulation in the Delta kept pace with a slow rise in sea level, forming thick deposits of peat capped by tidal marshes (Shlemon and Begg 1975; Atwater et al. 1979; Malamud-Roam et al. 2007). A century and a half of farming has reversed this process, creating artificial islands that are mostly below sea level, protected only by fragile levees (Drexler et al. 2007). Today, those who drive through the Delta see mainly huge tracts of flat, prosperous farmland intersected by narrow channels populated by recreational boaters.

Geographically, the area known as the "Legal Delta" lies roughly between the cities of Sacramento, Stockton, Tracy, and Antioch (p. xix). It extends approximately 24 miles east to west and 48 miles north to south and includes parts of five counties (Sacramento, San Joaquin, Contra Costa, Solano, and Yolo). At its western edge lies Suisun Marsh, an integral part of the Delta ecosystem. At its southern end, near Tracy, motorists pass over two major pieces of California's water infrastructure—the Delta-Mendota Canal and the California Aqueduct. These and several smaller aqueducts, built between the 1930s and the 1960s, deliver water from Northern California rivers to cities and farmland in coastal and Southern California and the San Joaquin Valley. The Delta is considered the hub of the state's

DELTA MISNOMERS

There is a long tradition in the Delta of improperly naming its physical features. At the top of the list is the term "Delta." Throughout the rest of the world (with a few exceptions, such as the Okavango Delta), deltas are formed where rivers disgorge into open bodies of water, leaving a prism of sediment, often of a shape similar to the Greek symbol Δ. The Sacramento–San Joaquin "Delta" does not qualify as a traditional delta, since it is formed at the tidally influenced confluence of two large floodplain rivers. The second and third most common misnomers are the terms "levees" and "islands." Levees are earthen embankments that hold back water during floods. The "levees" of the Delta are truly *dikes* that hold back water all the time. Similarly, islands are lands of positive relief surrounded by water. The Delta's "islands" are reclaimed lands that form topographic depressions surrounded by water. In this regard, they are *polders* instead of islands. Despite these differences, we use the terms "Delta," "levee," and "island" to match local convention in California. The Dutch, whose delta landscape employs many dikes to maintain their polders, have a different and more authoritative usage.

water supply because it is used as a transit point for this water. This role has significantly influenced Delta management policies, which aim to keep Delta water fresh.

Today, the Delta supports a highly modified ecosystem. It resembles the Delta of the past only in that some of the original species, such as delta smelt and Chinook salmon, are still present, albeit in diminished numbers. Invasive organisms, from plants to fish species, now dominate the Delta's steep-sided channels and long-flooded islands (mainly Franks Tract and Mildred Island).[1] Most of the native fish either migrate through the Delta (e.g., Chinook salmon, steelhead, and splittail) or move into it for spawning (delta smelt and longfin smelt). Resident native fish are present mainly in areas strongly influenced by Sacramento River flows. Recent years have seen spectacular declines in salmon populations (often called "anadromous" because they live in ocean water and move inland to spawn), the delta smelt, longfin smelt, and other open-water or "pelagic" species. Habitats in marshlands and along the banks of rivers ("riparian" areas) have been reduced to small remnants in the Delta, although agricultural lands are important winter foraging areas for sandhill cranes and various waterfowl (Herbold and Moyle 1989).

WHY THE DELTA MATTERS TO CALIFORNIANS

Most Californians rely on the Delta for something, whether they know it or not. About half of California's average annual streamflow flows toward the Delta. Most Californians drink water that passes through the Delta, and most of California's farmland depends on water tributary to the Delta.[2] In addition, increasingly, people are building homes in the Delta, perhaps not realizing the risks to their property and lives from living near or below sea level behind undersized and insufficiently maintained levees. Table 1.1 summarizes the many ways in which California's regions receive services from the Delta.

Clearly, the Delta is not merely a hub for water supply. It is also a center for important components of California's civil infrastructure (Figure 1.1). The electricity and gas transmission lines that crisscross the region serve many parts of the state. The Delta is also used for the underground storage of natural gas to accommodate peak wintertime demands. Furthermore, the Delta hosts several transportation lines. California's major north-south highway (I-5) goes through its eastern edge, and two commuter routes—SR 4 and SR 12—cross its southern and central portions, respectively (Figure 1.1). Several rail lines pass through the heart of the Delta, as do the deepwater ship channels leading to the ports of Stockton and Sacramento. In addition, aqueducts and canals conveying water to several west-of-Delta water utilities—including the East Bay Municipal Utilities District and the Contra Costa Water District—also pass through parts of the Delta. Two power plants are at the Delta's western edge in Antioch and Pittsburg.

As noted earlier, the Delta also provides crucial habitat: Many of California's fish species live in or migrate through it. Moreover, the Delta is valued for its aesthetic appeal and support of recreational activities. Its proximity to population centers in the Bay Area, Sacramento, and northern San Joaquin Valley makes it an attractive destination for boating, fishing, hunting, and ecotourism. The Delta's 635 miles of boating waterways are served by 95 marinas containing 11,700 in-water boat slips and dry storage for 5,500 boats. In 2000, there were an estimated 6.4 million boating-related visitor-days, with 2.13 million boating trips. Recreational boating is expected to grow to 8.0 million visitor-days by 2020 (Department of Boating and Waterways 2002). Fishing is also a popular activity (Plater and Wade 2002), as is duck hunting in the Suisun Marsh.

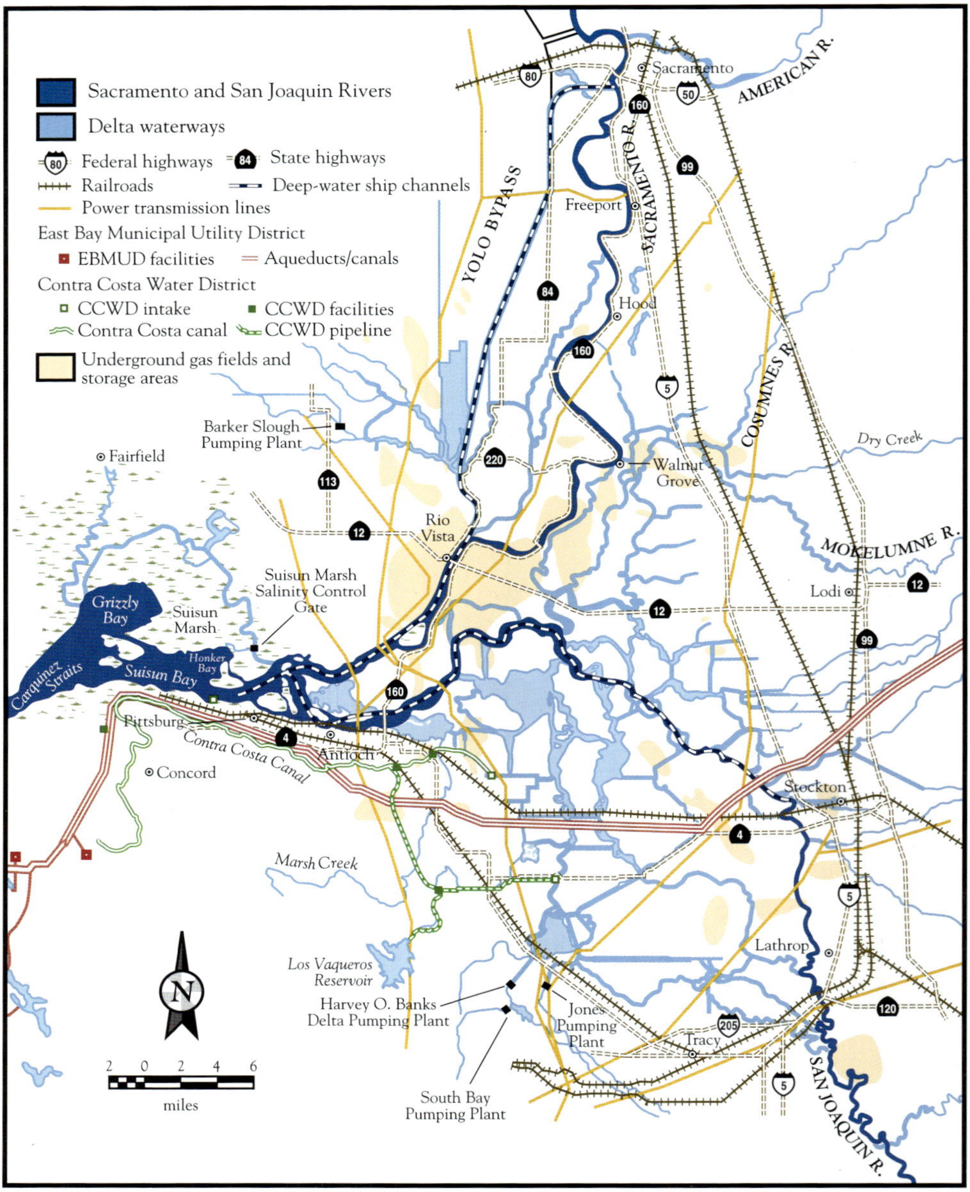

FIGURE 1.1 Infrastructure in the Delta.

TABLE 1.1 *Services Supplied by the Delta Region to Areas of California*

Delta Service	*Benefiting Region*			
	North of Delta[a]	In-Delta[b]	South of Delta[c]	West of Delta[d]
Agricultural land use		✓		
Urban land use		✓		
Ecosystem nutrients and support		✓		✓
Migration routes for salmon and other fish	✓	✓	✓	✓
Water supply	✓	✓	✓	✓
Recreation (boating, fishing, hunting, ecotourism)	✓	✓	✓	✓
Commercial shipping	✓	✓	✓	✓
Natural gas mining and power generation	✓	✓	✓	✓
Electricity and gas transmission and gas storage	✓	✓	✓	✓
Road and rail connections	✓	✓	✓	✓
Salt, waste, and drainage disposal	✓	✓	✓	
Water supply right-of-way				✓

[a]North of Delta includes the Sacramento Valley.

[b]In-Delta includes Delta islands.

[c]South of Delta includes Southern California and the eight-county San Joaquin Valley.

[d]West of Delta includes the San Francisco Bay Area (including Contra Costa County).

The Delta also serves as a vast drainage area for polluted agricultural and urban runoff. This runoff contains a variety of surplus and residual pesticides and nutrients, in addition to contaminants leached from the soils of upstream regions. Drainage from within the Delta contains dissolved organic compounds from the islands' peaty soils, which increase water treatment costs and drinking water quality risks. Sacramento Valley drainage includes mercury and other wastes from historic mining activities, and San Joaquin Valley agricultural drainage includes salts originating in the soils from its west side and sea salts and agricultural drainage intruding into irrigation supplies. Retaining such wastes locally would cause great expense and impairment within the source regions, but allowing them to flow into the Delta creates water quality problems for human and environmental uses within the Delta and beyond.

Finally, the Delta provides land. Until recently this land had been used predominantly for agriculture. Today, however, the Delta's land, as well as its water, has come into greater demand for urban, environmental, and recreational uses.

THE DELTA IN CRISIS

Concerns for the continued provision of services from the Delta involve several issues:

- Land subsidence, sea-level rise, and changes in climate make Delta levees increasingly vulnerable to failure from earthquakes, floods, and other causes.
- Endangered species and fisheries have continued to decline, and disruptive nonnative species continue to invade.
- Delta water quality remains at risk from salts entering from the ocean and the San Joaquin Valley's agricultural drainage, as well as from pesticides, metals, and other contaminants from agricultural and urban lands.
- Regional population and economic growth have increased pressure to urbanize Delta lands near major transportation routes and urban centers. This "hardening" of Delta lands simultaneously raises the costs of flood risks and reduces the flexibility of land management options.

Awareness of these issues has intensified in recent years, leading many to question the viability of current policies for the Delta. Indeed, by several key criteria, the Delta is now widely perceived to be in crisis. One aspect of the crisis is the health of the levees. The devastating effects of Hurricane Katrina on levees in New Orleans galvanized public attention on the fragility of the Delta's levee system, where close calls occur with some frequency; for example, a Jones Tract levee broke in June 2004. Recently, the Department of Water Resources (DWR) has publicized the economic consequences of a catastrophic levee failure caused by a large earthquake. One scenario, which envisaged 30 levee breaches and 16 flooded islands, predicted that water exports would be cut off for several months, that shipping to the

Port of Stockton would be cut off, and that there would be disruptions of power and road transportation lines (Snow 2006). The total cost to the economy over five years was estimated at $30 to $40 billion. A similar study of a 50-breach scenario that focused only on the costs to water users, put the annual cost of a shutdown of water exports at the pumps at $10 billion (Illingworth et al. 2005).

The second aspect of the crisis is the health of Delta fish species. In the fall of 2004, routine fish surveys registered sharp declines in several pelagic species, including the delta smelt, a species endemic to the Delta and listed as threatened under the Endangered Species Act (ESA). Subsequent surveys have confirmed the trend, raising concerns that the delta smelt—sometimes seen as an indicator of ecosystem health in the Delta—risks extinction if a solution is not found quickly (Figure 1.2).

The third aspect of the crisis is institutional. The CALFED process that has been responsible for coordinating Delta solutions since the mid-1990s has faced serious problems since late 2004. CALFED's failure to anticipate funding and disagreements among stakeholders on some key elements of its program has contributed to a loss of confidence in its institutional framework (Little Hoover Commission 2005). Since the summer of 2006, the California Bay Delta Authority—the body responsible for coordinating CALFED activities—has operated within the Natural Resources Agency, without an independent budget. Thus, the strong leadership and financial resources needed to address the Delta's problems are currently lacking. In this institutional vacuum, federal court decisions under the Endangered Species Act are now determining the conditions for water management. The state's recently completed Delta Vision process (Isenberg et al. 2008a, 2008b; Natural Resources Agency 2009) has signaled that state government and many stakeholders are aware that Delta policies are unsustainable and potentially catastrophic, and that they show a willingness to consider some major policy changes.

RESPONDING TO THE CRISIS

Recognition of the Delta crisis has led to appeals to pursue several very different management strategies. The collapse of Delta fish populations has prompted some environmentalists to call for cutbacks in water exports. Meanwhile, two main proposals have surfaced for dealing with levee

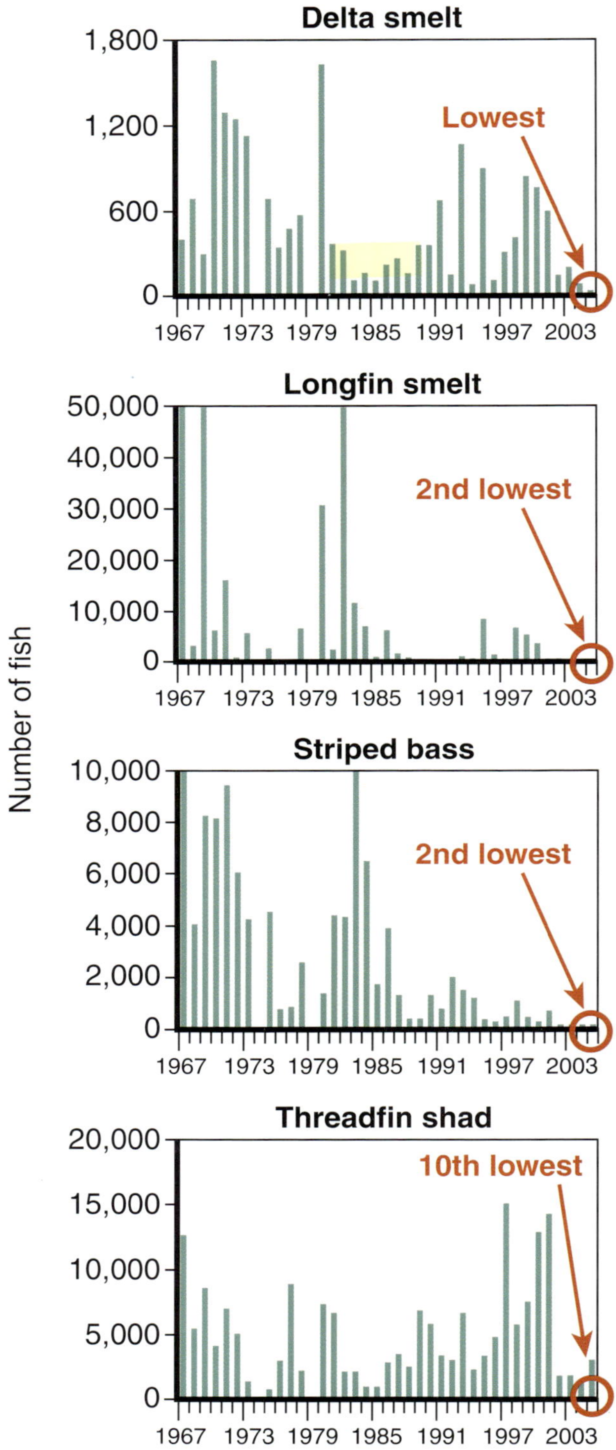
Delta smelt
Lowest
Longfin smelt
2nd lowest
Striped bass
2nd lowest
Threadfin shad
10th lowest
Number of fish
1967 1973 1979 1985 1991 1997 2003

FIGURE 1.2

instability: (1) Make massive investments in the levee system to reduce the risk of failure or (2) construct a peripheral canal around the Delta's eastern or western edge, to protect water exports from what many now view as unacceptable risks associated with direct Delta exports. The resurgence of a peripheral canal proposal is significant, because it is a solution that has deeply divided Californians in the past. Strong majorities of Northern California and San Joaquin Valley voters who were concerned over the canal's environmental effects, its potential to export too much water south, and the proposed allocation of costs, succeeded in defeating a peripheral canal proposal in a statewide referendum in 1982. When the CALFED process was launched in the mid-1990s to find new solutions to the Delta's ecosystem and water supply issues, feelings were still so raw that the peripheral canal was not considered an acceptable option.

These proposals have largely emerged from stakeholder groups, and none provides fully fleshed-out plans to address the Delta's woes. To date, the only concrete response from Sacramento, supported by both the governor and the legislature, has been to put more state funds into shoring up Delta levees, which were relatively neglected under CALFED.[3] State budget allocations for levee repairs were increased significantly in 2006, and two bond measures passed in the November 2006 ballot allocate additional funds for flood control in the Delta. However, there is as yet only the beginning of a broad plan for responding to the crisis in the Delta (Natural Resources Agency 2009).

The task at hand is urgent, and the stakes in the Delta are high. If California fails to develop a viable solution and act soon, the risk is loss of native species and significant disruptions of economic activity. Yet there is also a risk that the political process will choose expedient, incremental solutions that preclude or defer essential strategic decisions or prematurely close off the consideration of options that could help California make the most of the Delta, while protecting its unique ecosystem and species.

FIGURE 1.2 Fall abundance indices for several pelagic fish species in the Delta, 1967–2005. Graphs report the indices for the fall midwater trawl. Circles indicate the rank of indices in 2005. For delta smelt, longfin smelt, and striped bass, the recent indices represent low points in long-term declines of their populations. Source: California Department of Fish and Game.

FOUR CENTRAL ISSUES

Long-term solutions for the Delta will need to consider a wider range of issues than simply which levees to upgrade. To be viable, Delta solutions will need to address four central issues: the salinity and quality of Delta waters, in-Delta land use and water supply, water supply exports, and the Delta ecosystem.

DELTA SALINITY AND WATER QUALITY

With rivers feeding into it and marine bays at its western edge, the Delta is the meeting point for seawater and fresh water within the wider estuary system (Knowles 2002). Delta salinity has been a major concern since the city of Antioch's 1920 lawsuit against irrigators in the Sacramento Valley, whose upstream water withdrawals reduced freshwater flows into the Delta and increased the salinity at water intakes in the western Delta (Jackson and Paterson 1977). Salinity affects the potability and taste of urban water supplies, the productivity of farmland, and the viability of different organisms within aquatic ecosystems. For many decades, this issue was discussed in terms of where the salinity gradient—that is, the transition from fresh water to seawater—should be located in the estuary. Since the 1920s, it has been regarded as desirable to maintain the Delta, as much as possible, as a freshwater system, Suisun Bay and Marsh as brackish water systems, and San Francisco Bay as a marine (saltwater) system. The current regulatory framework for water quality in the Delta rests on this idea. More recent thinking, discussed in Chapter 6 and Moyle and Bennett (2008), holds that seasonal and interannual variability in much of the estuary may better mimic the natural salinity regime and help limit the extent of invasive species, which tend to prefer waters with little salinity fluctuation. Increasingly, it has been recognized that salinity and other, broader water quality problems in the Delta are compounded by the quality of upstream and in-Delta drainage, with consequences both for urban and agricultural users as well as for fish and wildlife.

DELTA LAND USE

Land is a central issue for the Delta. Of the Delta's 738,000 acres, roughly two-thirds support agriculture and one-tenth urbanized populations. Although the human population within the heart of the Delta is minimal—limited principally to homesteads and a handful of small "legacy" towns—larger cities such as Stockton, Antioch, and West Sacramento have long existed on its fringes. The Delta is often thought of as a site of high-value fruit and vegetable farms, but roughly 75 percent of the farmland is actually devoted to lower-value pasture and field crops; in comparison, only 55 percent of farmland statewide is devoted to these uses (Department of Water Resources 1998). And in recent years, urbanization and recreational use of Delta lands have been on the rise

Various environmental uses of Delta land already exist, including wetlands, riparian habitat, waterfowl uses, and aquatic habitats. Open water—which results when islands are flooded and submerged—also has environmental use, as well as considerable value for recreation, boating, and shipping. Freshwater storage is another recent suggestion for Delta lands. This freshwater storage plan proposes investing in strengthening internal levees on some Delta islands that have subsided below sea level, allowing them to be filled with water, on a tidal or seasonal time scale, to aid water projects in pumping fresh water from the Delta.

Each of these land uses has different implications for water use, the quality of water required in adjacent channels, drainage quality and quantity, and economic sustainability. Fortunately, the Delta is large and diverse enough to support a mix of land uses and habitats.

WATER EXPORTS

Water exports from the Delta are a major cause of controversy. For water users in the Bay Area, Southern California, and the San Joaquin Valley, the reliability and quality of these water supplies are of paramount concern. Yet there are also concerns that export patterns and volumes harm species' health and water quality within the Delta. Many approaches exist for either providing or avoiding this function for the Delta, and numerous options have been proposed over the past century. Even without providing water exports, however, the Delta would still have many serious problems with flooding, land subsidence, degraded habitat, invasive species, and water quality.

DELTA ECOSYSTEM

Different parts of the Delta provide habitat for different wild species and their diverse life stages. The mix of salt, brackish, and freshwater marshes as well as upland, riverine, and deepwater habitats affects the abundance and distribution of native and alien species. Therefore, anything that changes the physical Delta changes the biological Delta. Since the 1970s, considerable attention has been paid to the effect of water supply functions on ecosystem functions in the Delta. Initially, this discussion focused primarily on the role of water export pumps at the Delta's southern edge, and on efforts to avoid fish entrainment (the drawing and trapping of fish into the pumps). It is now recognized that the same issues of entrainment of fish and invertebrates apply to power plant cooling water and agricultural and urban diversions elsewhere in the Delta. Concerns have also been raised that the total volume and timing of diversions are causing problems for key Delta species by changing the way water flows through the Delta. Given the range of federal and state environmental laws protecting these species, these concerns are legal and political as much as ecological (National Marine Fisheries Service 2008).

SEARCHING FOR A SOFT LANDING

In this book, we look for long-term solutions to these chronic, dire, and potentially catastrophic problems. Rather than focus on crisis management, we consider long-term strategies, under which Californians can develop and implement a plan to adjust to the Delta of the future. This approach, which we refer to as planning for a "soft landing," differs greatly from how California may need to manage short-term crises in the Delta, or what might be considered a "hard landing." If the state is unfortunate enough to experience a multi-levee failure before implementing a long-term plan, effective emergency response will be needed to minimize the costs in terms of water supply and damages to other economic infrastructure.

In assessing long-term solutions, we consider the physical factors that will force a major transition on the Delta, including climate change, sea-level rise, earthquakes, and changing flood flows. We unite perspectives from a wide range of disciplines that are important to Delta analysis—engineering, biology, geology, and economics. We focus on two central questions for long-term Delta policy. First, which Delta islands should be repaired when they fail? We consider the economic sustainability of investment strategies in Delta levees, given the risks of failure, the value of land and other assets, and the costs of protecting Delta lands. Second, what is the preferable long-term water management strategy from a statewide perspective: continued through-Delta export pumping, a peripheral canal, a dual conveyance system combining through-Delta pumping and a peripheral canal, or ending exports altogether? We compare futures for the ecosystem and California's economy under these four broad water export alternatives, which span the full range of potential strategies. Our aim is to advance policy and public discussions about Delta futures. These broad alternatives present choices that must be made *before* addressing the many decisions required to implement any chosen strategy.

Quantitative risk analysis is the central framework used to evaluate performance, integrating estimates of costs and probabilities with major uncertainties in the performance of each alternative. This framework is particularly useful for assessing policy alternatives in the Delta, given the risks and uncertainties for ecosystem outcomes as well as changing physical conditions in the Delta. To conduct such a risk analysis required detailed analysis of levee risks and economics, hydrodynamics and water quality under future conditions of climate change, analysis of ecosystem response to changes in the Delta, and economic analysis of how California's water supply system could respond to major changes in Delta water export policies. Along the way, we also seek to provide insights into implementation and the governance, finance, and regulatory changes needed to improve the prospects of the Delta from environmental and statewide economic perspectives. Greater detail on many of these topics is provided in a series of appendices to Lund et al. (2008a). All cost estimates are presented in 2008 dollars.

Our analysis does not provide perfect clarity, but perfect clarity should not be needed to select a strategy to solve an urgent problem. We come to firmer and better substantiated conclusions than we expected regarding both Delta island repair and strategic directions for Delta water exports. These findings provide building blocks for a promising new approach to managing this unique and challenged resource, as Californians struggle to preserve the many ecological, economic, and cultural functions of the Delta in the face of inevitable changes.

2

THE LEGACIES OF DELTA HISTORY

> You could not step twice into the same river; for other waters are ever flowing on to you.
>
> HERACLITUS (540 BC–480 BC)

The modern history of the Delta reveals profound geologic and social changes that began with European settlement in the mid-nineteenth century. After 1800, the Delta evolved from a fishing, hunting, and foraging site for Native Americans (primarily Miwok and Wintun tribes), to a transportation network for explorers and settlers, to a major agrarian resource for California, and finally to the hub of the water supply system for San Joaquin Valley agriculture and Southern California and Bay Area cities. Central to these transformations was the conversion of vast areas of tidal wetlands into islands of farmland surrounded by levees. Much like the history of the Florida Everglades (Grunwald 2006), each transformation was made without the benefit of knowing future needs and uses; collectively these changes have brought the Delta to its current state.

PRE-EUROPEAN DELTA: FLUCTUATING SALINITY AND LANDS

As originally found by European explorers, nearly 60 percent of the Delta was submerged by daily tides, and spring tides could submerge it entirely.[1] Large areas were also subject to seasonal river flooding. Although most of the Delta was a tidal wetland, the water within the interior remained predominantly fresh. However, early explorers reported evidence of saltwater intrusion during the summer months in some years (Jackson and Paterson 1977). Dominant vegetation included tules—marsh plants that live in fresh and brackish water. On higher ground, including the numerous natural levees formed by silt deposits, plant life consisted of coarse grasses; willows; blackberry and wild rose thickets; and galleries of oak, sycamore, alder, walnut, and cottonwood. Few traces of this earlier plant life remain; agricultural practices and urbanization have cleared most forested areas and levee upgrading has removed most trees and vegetation from previously natural levees.

Before European settlement, the Delta also teemed with game animals and birds. Elk, deer, antelope, and grizzly bear frequented the tules and the more open countryside. Sightings of elk were reported as late as 1874, but the last of the large game animals were likely destroyed by the 1878 flood.

From the reports of early explorers, it has been estimated that the native population in the Delta area was between 3,000 and 15,000. Most native villages were on natural levees on the edges of the eastern Delta and typically contained around 200 residents, although one community was thought to contain at least 1,000 residents. The native population did not practice agriculture, although they did manage the landscape with fire and other tools to favor plants they used (Anderson 2005). Their diet consisted of the roots and pollen of the tules, acorns, and the fruit and seeds of other wild plants. Fish and game were also important staples.

European settlement of the Delta began slowly. Despite several expeditions between 1806 and 1812, the Spanish failed to locate a suitable site for missions in the region. From 1813 to 1845, most expeditions were military attempts to subdue the native population. The Hudson Bay Company sent trappers into the Delta from 1828 through 1843, but had limited success because of interference by Native Americans, priests, and local merchants. From 1835 through 1846, the Spanish established several land grants. In 1841, John Sutter was the first foreigner to be granted land in the Delta vicinity. By 1846, an estimated 150 European-Americans were in the Central Valley, mostly at Sutter's Fort near present-day Sacramento. A Dutchman living on an unconfirmed grant below Sutter's Landing was the only certain European-American resident within the Delta, with others scattered on the periphery.

Two events in 1847 set the stage for accelerated settlement of the Delta. The first was the transfer of California to the United States at the end of the Mexican-American war; many U.S. soldiers had volunteered for the war with the idea of staying in California. The second was the introduction of the steamboat, *Sutter's Sitka*. The *Sitka* reduced travel time from Sacramento to San Francisco from a typical two- to three-week trip to just under seven days, a change that greatly facilitated trade throughout the Delta.

RECLAMATION: FOUNDATIONS OF THE MODERN DELTA ECONOMY

The reclamation of Delta lands began almost simultaneously with the California gold rush. Within weeks of the January 1848 discovery, the few settlements near the coast had all but emptied, and an influx of tens of thousands of people followed. Almost immediately, many miners saw surer fortunes to be made from tilling the soil than from mining. Most of them selected lands on the natural levees of the main waterways or on higher ground near streams close to heavily traveled trails. By the early 1850s, interest turned to the diking and draining of flooded Delta lands.

The reclamation era spanned over 80 years and was marked by frequent institutional change, as Delta interests and state and federal authorities sought to tackle problems ranging from basic levee construction, to regional flood control and maintenance of shipping channels, to salinity intrusion. Many of these problems were compounded by the presence of upstream mining activities, which sent massive volumes of debris into the Delta. Although most land reclamation was undertaken by private individuals or local groups, this era witnessed the first major public works project in the Delta—the Central Valley flood control system. By the time the last Delta island was diked and drained in the early 1930s, Delta farmers and the cities on the Delta's periphery had become firmly established interests whose concerns over water quality would figure prominently in the search for large-scale solutions to Delta water issues in subsequent decades.

RECLAMATION AND THE RISE OF DELTA AGRICULTURE

Delta reclamation is a process that becomes increasingly difficult as it progresses. Each acre of drained and diked land represents the removal of floodplains, placing more stress on the remaining system by reducing space for subsequent floodwaters to occupy. Initial reclamation efforts amounted to little more than attempts to supplement natural levees to protect agricultural plots during high tides and seasonal floods. It soon became clear that for reclamation to proceed, institutions were needed to provide land tenure security and to facilitate collective work on levees.

A primary piece of enabling legislation for the reclamation of Delta lands was the Arkansas Act of 1850, more commonly known as the Swampland Act. This law ceded federal swamplands to the states to encourage their reclamation. California received 2,192,506 acres, including nearly 500,000 acres within the Delta. Sales began in 1858. Initially, individual acquisitions were limited to 320 acres, at the price of $1 per acre (about $23 per acre in today's dollars). In 1859, the size limit was doubled to 640 acres, and limits were repealed altogether in 1868.

Although several continuous levees were built in the 1850s (notably, on Grand and Sherman islands), collective levee building was facilitated by the creation of the State Board of Reclamation in 1861, which was given the authority to form reclamation districts from collectives of smaller parcel owners (see map Delta Islands, on page xx, for the location of individual islands). Between 1861 and 1866, the board authorized reclamation districts to enclose large areas that were defined by natural levees. The board also embarked on several large-scale schemes to reclaim lands and provide flood protection in the Sacramento and Yolo basins and on several Delta islands. Although the board was dissolved before much of this work could be completed, its duties were transferred to the counties, which continued to oversee the creation of reclamation and levee maintenance districts. Ninety-three of these local agencies still operate within the Delta today, with frontline responsibility for levee maintenance.

Technology also played a central role in reclamation. A contractor in charge of levee construction on Staten Island, J. T. Bailey, developed the first mechanized equipment for levee construction in 1865 (Thompson 1957). After 1868, when the 640-acre size limit was repealed, corporate speculators and wealthy individuals undertook large-scale reclamation and derived profits from selling the improved land. Machine power was applied to levee construction, land clearing, ditch building, and dredging, and pumps were introduced to drain the parcels.

The influence of these institutional and technological innovations on the pace of reclamation is striking (Table 2.1). In the 1870s, over 90,000 acres were reclaimed, six times more than in the preceding decade. Reclamation

TABLE 2.1 *Reclamation Growth in the Delta*

Decade	*Acres Reclaimed*	*Cumulative Acres*
1860–1870	15,000	15,000
1870–1880	92,000	107,000
1880–1890	70,000	177,000
1890–1900	58,000	235,000
1900–1910	88,000	323,000
1910–1920	94,000	417,000
1920–1930	24,000	441,000

SOURCE: Thompson 1957.

efforts in the Delta continued through the 1930s, with the last island, McCormack-Williamson Tract, reclaimed in 1934.

In the early years of reclamation, the Delta was seen as a drought-free, fertile area on which the state could depend to support its growth. Delta waterways provided natural and inexpensive transportation routes. The droughts that ruined San Joaquin Valley wheat and barley crops served to further enhance the value of Delta farmlands. An editorial in the San Francisco *Alta* of July 25, 1869, provides a characteristic view:

> In these reclaimable lands we shall have drought-proof means of life and luxurious living for the whole population of our State, were it twice as numerous. Heretofore the certainty of occasional famine years has been a dark cloud on the horizon before the thoughtful vision. Now we see salvation. All hail! to the great minds that have conceived this enterprise. God speed their success and bring them rich reward.

These high hopes waned after the major floods of 1878 and 1881, which revealed the susceptibility of reclaimed lands to recurrent inundations. By this time, however, Delta agriculture had become an important interest in its own right, with landowners seeking relief from floods and mining debris (and, eventually, from salinity intrusion) through judicial and political channels.

LEGAL BATTLES OVER UPSTREAM MINING

It is estimated that between 1860 and 1914, more than 800 million cubic yards of mining debris—enough to fill 10,000 football fields to a depth of 16 yards—passed through the Delta, primarily from hydraulic mining sites upstream of the Sacramento River watersheds. Although this debris had some positive effects—notably by bolstering levees and providing fill material—its overall consequences were decidedly negative. The debris raised and constricted the channels, worsening the reduced tidal action caused by reclamation. Consequences included transportation difficulties, increased susceptibility to flooding, and decreased agricultural productivity. (The latter problem, a result of seepage from an elevated water table, was mitigated somewhat when pumps became available in the early 1900s.)

In 1880, the state legislature formed the Board of Drainage Commissioners in an attempt to find a solution between the miners and the farmers. The board was to create drainage basin planning districts with the costs borne by a statewide land tax and taxes on hydraulic mining. When the State Supreme Court invalidated this action the following year, farmers instituted injunction proceedings against the miners. The first of these cases—*People v. Gold Run Ditch and Mining Company* (July 1881)—is considered a landmark piece of environmental jurisprudence. It invoked the public trust doctrine to impose an injunction on hydraulic mining. A second case, *Woodruff v. North Bloomfield Gravel Company* (January 1884), also sided with the farmers.

PUBLIC WORKS FOR FLOOD CONTROL

In reaction to these rulings and to pressure from Central Valley business interests, subsequent decades saw a flurry of attempts to find a comprehensive solution to flooding issues in the Delta and the greater watersheds of the Sacramento and San Joaquin rivers. The result was a series of major public investments, involving both the federal and state governments, which are still core elements of the Central Valley flood control system.

The 1893 Caminetti Act authorized the federal government to cooperate with California in formulating plans to prevent mining tailings from passing downstream. The California Debris Commission—a three-member body of army engineers—was created to work with the federal government in this effort. Although the commission's primary goal was to find a way to resume mining without the tailings problem, its legacy was regional flood control (Kelley 1989). In 1910, the commission initiated dredging of the lower Sacramento River, under what was known as the "Minor Project."[2] A commission report submitted to Congress in 1911 formed the basis of a comprehensive flood control plan for the Sacramento River. This expanded plan (dubbed the "Major Project") included proposals for continued channel dredging and the creation of the Yolo Bypass, which provides floodway space on private farmlands.[3] The plan also specified levee heights throughout the Delta.

When California's legislature approved the Major Project in 1911, it also resumed control over reclamation authority, recreating the Board of Reclamation to coordinate state reclamation, flood control, and navigation improvement. The U.S. Congress approved the Major Project in 1917, after the state and landowners agreed to greater participation. The Federal Flood Control Act of 1928 grew from the California Debris Commission's study (as well as Mississippi River experiences) and marked congressional recognition of responsibility in flood control as well as navigation.

Today, flood control within the Central Valley continues to operate under this system of joint responsibility. Federal and state agencies have the primary charge for maintaining roughly 1,600 miles of publicly owned "project levees." Some cost-sharing of project levees is assumed by local reclamation districts and flood control agencies. Within the Delta itself, the mix of responsibilities is more complex. The Delta contains nearly 400 miles of project levees (notably the levees protecting the cities of Lathrop and Stockton) and over 700 miles of nonproject (sometimes called "private") agricultural levees, which are owned and operated by local reclamation districts and have limited state cost-sharing (Figure 2.1). Concerns have recently arisen regarding many aspects of the Central Valley flood control system, including the condition of project levees surrounding Sacramento and other upstream locations, but the nonproject Delta levees are a particularly weak link in the system.

THE EXPANSION OF SHIPPING CHANNELS

In the early twentieth century, the U.S. Army Corps of Engineers also became active in maintaining and improving shipping channels, which had suffered from debris buildup. The earliest efforts focused on the Sacramento corridor. From 1899 to 1927, the corps maintained a channel 7 feet deep between Suisun Bay and Sacramento; it was subsequently deepened to 10 feet. In 1946, Congress authorized a project to convert Sacramento into a deepwater port; the dredging of the 30-foot-deep channel was completed

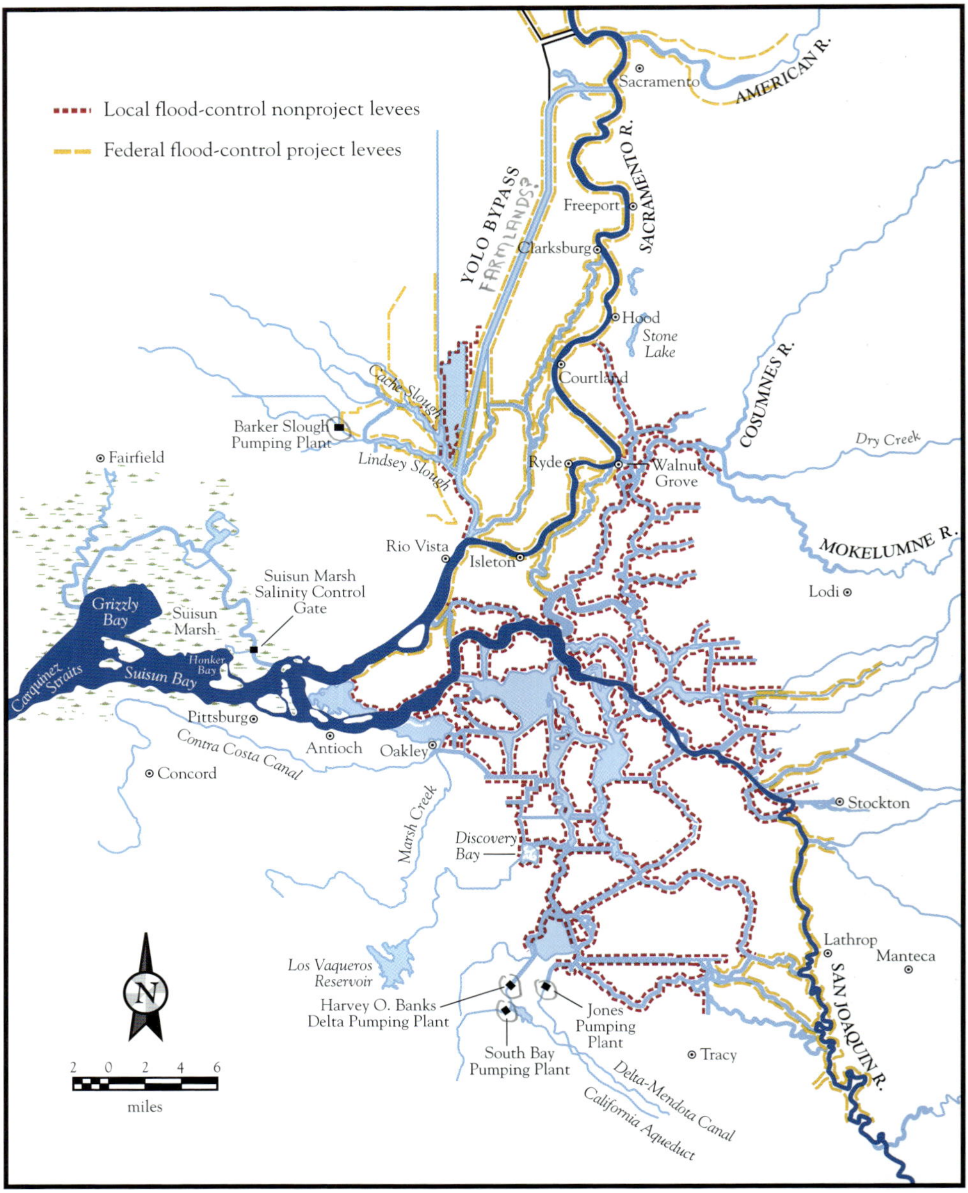

FIGURE 2.1 Delta levees, 2006.

in 1955. Similar efforts took place to improve shipping to the eastern Delta. The Stockton channel on the San Joaquin River was maintained at 9 feet from 1913 to 1933 and then dredged to 26 feet. In 1950 it was dredged to 30 feet, and in 1987 it was dredged to its current depth of 37 feet at low tide.

These deepwater shipping channels have altered water flows within the Delta.[4] As a result of dredging, water moves much more slowly through the lower Sacramento River than it does in shallower parts of the Delta, thereby providing a different environment for fish and other aquatic life. The Stockton ship channel is particularly important for east-west tidal exchange with the western Delta. Both the Sacramento and the Stockton ship channels (particularly the Stockton channel) would be threatened by a catastrophic levee failure, which could reintroduce large quantities of sediment into them. At present, these ports are relatively minor players in California's sea trade, although Stockton handles large volumes of agricultural produce from the Central Valley.[5] Sacramento traffic is anticipated to increase under a new management arrangement with the Port of Oakland (Port of Sacramento 2006).

THE FIRST SALINITY LAWSUITS

By the early twentieth century, salinity intrusion had become a major concern for Delta interests. Although it is not certain how far upstream ocean salinity extended under natural conditions, salinity levels did not hamper reclamation in the Delta as they did around the San Francisco Bay (Jackson and Paterson 1977). In the Delta, virgin reclaimed tracts did not need salts flushed out before agricultural practices began. In this period, salinity intrusion was seasonally highest in the late summer months after the mountain snowpack had melted, and salt water reached farther inland during very dry years, such as 1871 (Young 1929). However, the reduction of tidal floodplains through reclamation and mining debris deposits decreased the penetration of salt into the Delta (Matthew 1931a). But upstream diversions for irrigation in the Sacramento Valley greatly increased salt intrusion during summer months, especially in dry years. As early as 1908, the sugar refinery at Crockett sent barges as far as 28 miles inland (well into the Delta) to gather fresh water during the dry season (Figure 2.2). During the drought years in the 1920s, salt water reached so far into the Delta that these barges were sent west to Marin instead of east into the Delta.

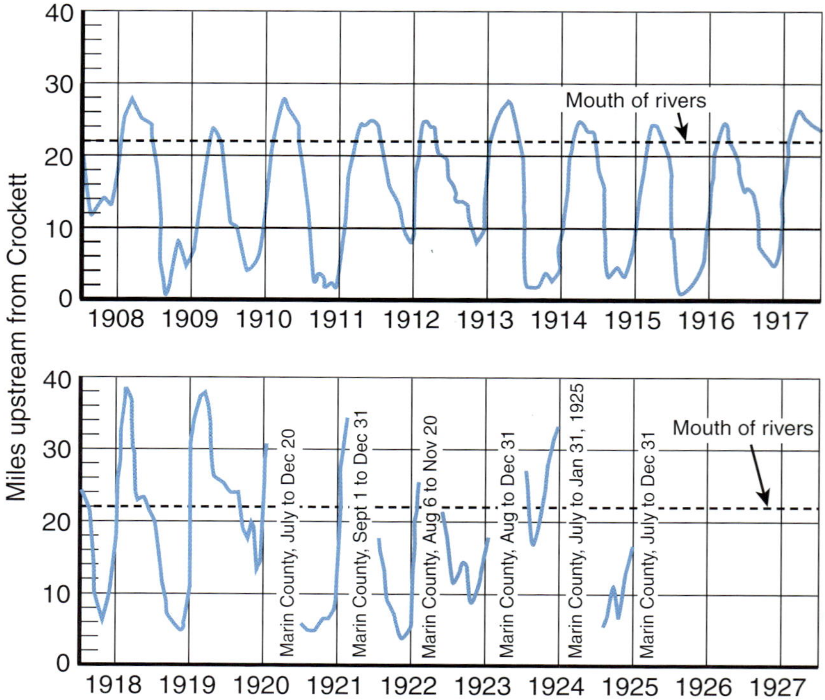

FIGURE 2.2 Upstream distance for barges looking for fresh water for sugar refinery at Crockett. Source: Young 1929, Plate 9-1.

Salt intrusion in the Delta reached its peak between 1910 and 1940, setting the stage for legal proceedings and various engineering proposals to keep the Delta fresh that have continued to this day.

The first salinity lawsuit was filed in July 1920 by the City of Antioch. The city, backed by various Delta interests, charged that upstream irrigators on the Sacramento River were diverting too much water, resulting in insufficient freshwater flows past Antioch to hold back ocean water.[6] Although the lower court initially ruled in Antioch's favor, the California Supreme Court overturned the decision on what is essentially a judgment of statewide costs versus benefits: To provide consistently fresh flows for the small community of Antioch at the Delta's western edge, much larger diversions for upstream irrigation would have had to be curtailed, an outcome the court considered unnecessarily wasteful.[7]

The suit nevertheless sparked efforts to find engineering solutions to the salinity problem. Initial proposals focused on the construction of a saltwater barrier in the outer part of the estuary, near the Carquinez Strait. A report from the state Department of Public Works (1923) officially endorsed this idea, which had already been considered on several occasions in the second half of the nineteenth century as a way to control floodwaters and to resolve rail transportation problems across the Delta (Jackson and Paterson 1977). Further support for a barrier came from those concerned about the effects of an invasive pest, the marine borer *Teredo navalis*, on docks and other wooden structures in the inland ports. This pest, one of the San Francisco Estuary's first invasive species, was moving upstream with salinity incursions. In the end, however, concerns over the high financial costs of a saltwater barrier, as well as the potential harm such a barrier would cause to commercial fisheries, led to its abandonment. Instead, as described below, control of Delta salinity was woven into projects to augment water supplies for users south of the Delta.

FARMING AND LAND SUBSIDENCE

Another problem that increased in severity over time was the subsidence of Delta lands, many of which now lie well below sea level (Figure 2.3). Reclamation itself initiated the subsidence process, because much of the material used to elevate the levees was taken from the interior of reclaimed islands, thereby lowering the island while elevating its protective barrier. Soil burning, mostly associated with the potato farming that developed by 1900, also accounted for much early subsidence. Despite the benefits of burning—weed control, fertilization, and the facilitation of the seedbed—it accelerated subsidence and allowed for salt accumulation and increased wind erosion.

Subsidence added to farming costs because it required additional levee rebuilding, drainage excavation, and pumping both for regular operations and recovery after floods. One casualty of this process was Franks Tract, which was abandoned and left flooded after a 1938 levee failure. The same fate befell Mildred Island in 1983 and Liberty Island in 1998. However, in general, Delta farmers have continued to farm subsided lands. As we describe in Chapter 3, even though the pace of subsidence has slowed in recent times, in part because some of the more destructive farming practices have ceased, subsidence of Delta islands continues and is a major contributor to levee instability.[8]

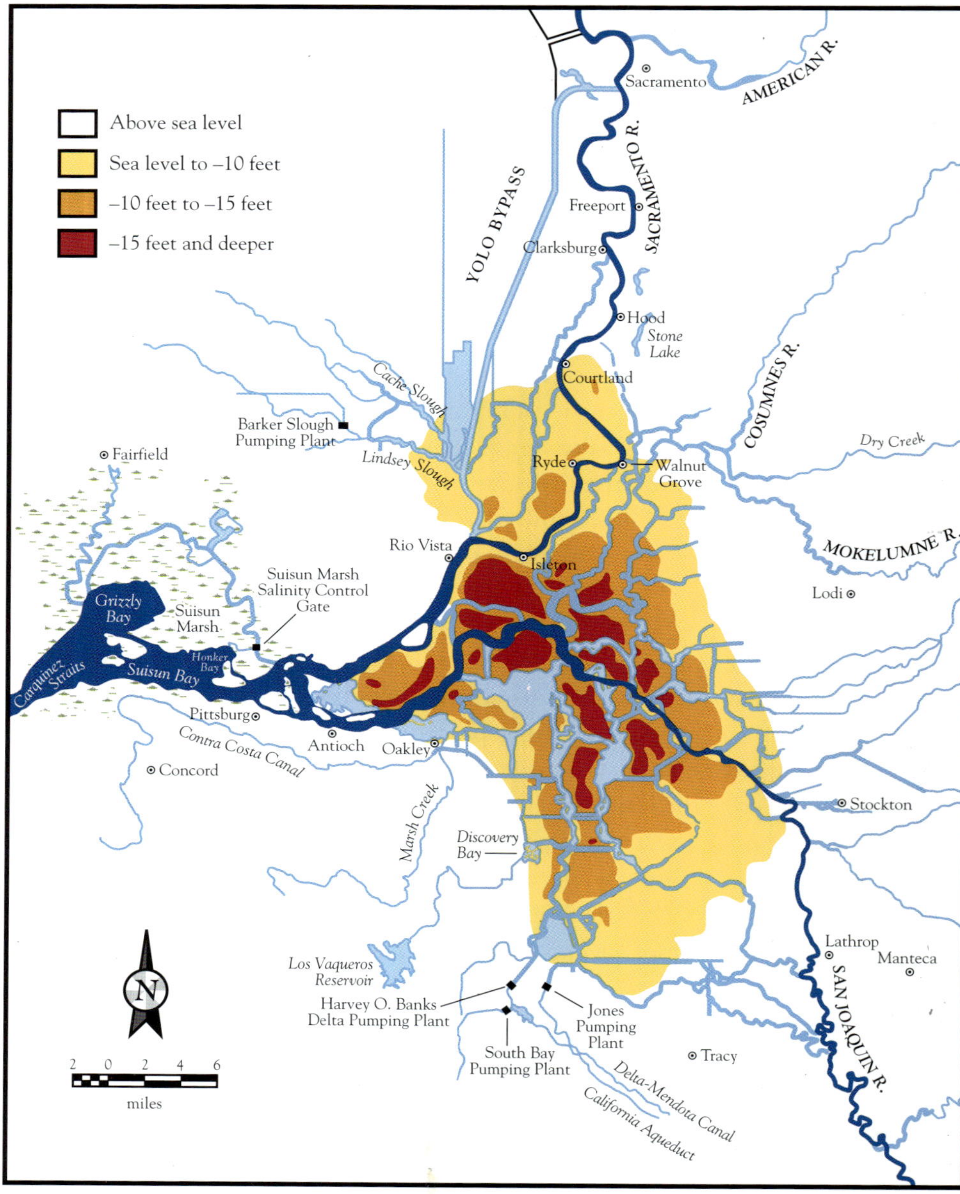

FIGURE 2.3 Land subsidence in the Delta. Source: Department of Water Resources 1995.

BIG WATER PROJECTS TRANSFORM THE DELTA TO A FRESHWATER BODY

By the time reclamation of Delta lands was nearly complete in the 1920s, attention began to focus on the development of water supplies from the two major Delta watersheds, the Sacramento and San Joaquin rivers. Elsewhere in California, major public works projects designed to move water across long distances had already been planned or undertaken, including the Los Angeles Aqueduct (from the Owens Valley to Los Angeles), the Hetch Hetchy project (bringing Sierra Nevada water to San Francisco), the Mokelumne River project (bringing Sierra Nevada water to the East Bay), and the investments along the Colorado River to deliver water to Southern California (Hundley 2001). From the 1930s to the early 1970s, the Central Valley witnessed a series of major investments in water storage and conveyance to supply agricultural and urban users. This process began with the federally sponsored Central Valley Project (CVP) and ended with the state-run State Water Project (SWP) and included some locally sponsored projects. Although some of the engineering analyses considered alternatives that bypassed the Delta, most of the investments actually undertaken relied on the Delta as a conduit for exports to points south and west (Jackson and Paterson 1977). Big water projects in the Delta have always generated debate, and many plans have been created, modified, and discarded. If nothing else, this process underscores the difficulties of managing the Delta—in the past as well as today.

THE CENTRAL VALLEY PROJECT

Since the late nineteenth century, various observers have recognized the potential for moving surplus Sacramento River water to the drier but potentially productive San Joaquin Valley (Alexander et al. 1874). The 1923 Department of Public Works' report to the legislature noted above included proposals to build upstream storage reservoirs to permit such transfers. These plans were fleshed out in the department's 1930 *State Water Plan* ("the Plan"), which would serve as a blueprint for the Central Valley Project (Department of Public Works 1930). The Plan concluded that upstream storage along the Sacramento River could simultaneously resolve two principal water problems: Water shortages in the San Joaquin Valley, where groundwater overdraft—or pumping in excess of natural recharge—had become a serious concern, and salinity intrusion in the

Delta, which would be addressed by creating a hydraulic salinity barrier, with controlled releases of water from upstream storage. Ultimately, the Plan rejected the idea of a physical salinity barrier, arguing that its construction could be postponed until the anticipated growth in San Joaquin Valley water demand used up excess reservoir water.[9] Salinity problems in the East Bay would be resolved by piping Delta supplies via a proposed Contra Costa County conduit. Investments along the Colorado River, meanwhile, were seen as the near-term solution to Southern California's additional water needs.

The legislature and voters approved the Central Valley Project in 1933. Seeking to maximize federal financial contributions in the hard economic times of the Depression, the state handed over control of the project to the federal government. Although construction of one of the CVP's primary components, Shasta Dam, got under way by 1938, state and federal agencies did not agree on the final form of diversions for Sacramento River water until the following decade. The U.S. Bureau of Reclamation (USBR) had proposed a new canal to route the water around the periphery of the Delta between Freeport and the Stockton area. The final outcome, closer to the state's original proposal, was to divert water through the Delta via a small cross-channel just north of Walnut Grove, from which it would travel south to the pumps. The Delta Cross Channel, constructed by USBR in 1944, still helps to supply the Contra Costa and Delta-Mendota canals, which entered service in 1948 and 1951, respectively.

The CVP has also been responsible for some major upstream diversions of water from both the Sacramento and San Joaquin rivers. Following the construction of Friant Dam (1942) and Friant-Kern Canal (1948), the CVP began diverting San Joaquin River water to supply irrigators on the east side of the San Joaquin Valley. Subsequent investments on the west side of the Sacramento Valley, notably the Tehama-Colusa Canal (1980), also increased upstream diversions from the Sacramento River.

The CVP was successful in its primary goals: expelling salt water from the Delta using controlled releases from Shasta Reservoir and supplying fresh water to irrigators and some urban users in the San Joaquin Valley and areas west of the Delta. The project also provided benefits to power generation and navigation. However, it was less successful in providing additional flood protection in the Delta, which continued to experience levee failures. Moreover, the CVP investments in water supply and salinity control were not considered adequate over the long run, given the anticipated growth in demand for water exports. Since the 1940s, a series of investigations have explored longer-term solutions to these issues. Salinity management in the Delta remains a major issue for the CVP.

THE STATE WATER PROJECT

In 1960, California voters authorized the first phase of the State Water Project, which aimed to extend water deliveries from northern watersheds to Southern California and Bay Area cities and to farmers in the Tulare Basin who were beyond the reach of the CVP. Although this project ultimately adopted the same basic approach to water exports as the CVP, relying on the Delta as a transfer point, this approach was not a foregone conclusion. Options that surfaced (or resurfaced) included a saltwater barrier, a highly reengineered and simplified Delta, and a peripheral canal. Investigations into the first two options took place in the 1950s. Peripheral canal investigations continued well into the 1970s, as part of the consideration of the SWP's expansion.

The foundation of the State Water Project was laid in the 1950s, through a series of proposals, plans, and legislative actions. In 1953, the state legislature passed the Abshire-Kelly Salinity Control Barrier Act to reexamine the need for a saltwater barrier. The state Division of Water Resources hired a Dutch consultant, Cornelius Biemond, who was Director of Water Supply for Metropolitan Amsterdam. Biemond rejected the idea of a barrier, proposing instead to reduce the Delta's 1,100 miles of levees to a 450-mile system of master levees. This plan included the construction of both a siphon to take Sacramento River water under the San Joaquin River on its way south and a barrier at the confluence of these two rivers.

By 1957, the newly formed Department of Water Resources (DWR) discarded the concept of a saltwater barrier in favor of a somewhat modified Biemond Plan and recommended it to the governor and legislature as part of the State Water Project (Department of Water Resources 1957). Under this proposal, water would be transferred through both a trans-Delta system (the Biemond Plan) and an Antioch Crossing Canal, along the Delta's western edge. Three pumping plants in the south Delta near Tracy would pump supplies farther southward. The Biemond Plan would isolate many Delta channels from tidal action, allowing salinity to be controlled with one-third of the available freshwater flow. In 1959, the Water

Resources Development Act was passed to pay for the first phase of the SWP and approved by voters in 1960.

Perhaps reflecting the growing political savvy of Delta interests, the SWP ran into greater public acceptance obstacles than had the CVP. As a precondition to the SWP's advancement, the legislature passed the Delta Protection Act of 1959, which established the legal geographical boundaries of the Delta and stipulated that the state-run SWP, in coordination with the federally run CVP, would be required to maintain Delta water quality standards (i.e., sufficiently low salinity to permit farming and other economic uses). However, Delta interests remained concerned about water quality, and in 1961, the State Assembly Interim Committee of Water rejected the Biemond Plan, stating that it was an imposed solution rather than one worked out in consultation with local interests.

While work began on the SWP's main storage and conveyance components—Oroville Dam and the California Aqueduct—deliberations continued on the ultimate solution for moving water from north to south. An Interagency Delta Committee was formed to examine Delta water problems. As one alternative, USBR revised the peripheral canal proposal from the 1940s.[10] The committee also examined options for keeping the entire Delta fresh, either with a physical barrier at Chipps Island on the Delta's western edge or through the continued use of controlled reservoir releases to maintain a hydraulic saltwater barrier.

In 1964, the committee released its *Proposed Report on Plan of Development, Sacramento–San Joaquin Delta*, again recommending the peripheral canal but with several refinements, including an increase in the volume of diversions from the Sacramento River to supply south-of-Delta users. The report stressed the intangible environmental benefits of the canal and proposed further work to safeguard the water supplies of western counties. In public hearings, only Contra Costa County raised objections to the canal proposal, while environmental groups remained supportive.

The peripheral canal was on its way to becoming a reality. By 1966, DWR had officially adopted the canal as a part of the State Water Project and had reached agreements on cost-sharing provisions with USBR. Public meetings were held to gather local input on proposed canal alignments. While waiting for congressional authorization, the new director of DWR placed the project design on hold but continued with right-of-way purchases. In 1969, USBR released its economic feasibility study and recommended that Congress approve the project. Both chambers of the California legislature issued strong endorsements of the canal. But despite its promising start, this version of the peripheral canal never came to be—other forces were at work that changed the course of the debate about the Delta.

ENVIRONMENTAL CONCERNS CHANGE THE COURSE OF DELTA POLICY DEBATES

The SWP's plans would all change over the following decade, as California, like the nation as a whole, witnessed the rise of environmental concerns. This shift in public attitudes was reflected in new legal and regulatory frameworks for pollution control and species protection. The Delta and its tributary watersheds, home to many unique aquatic species, would become a focal point for these new concerns. One casualty would be the build-out of the State Water Project, as northern rivers slated as sources for additional upstream storage were declared "Wild and Scenic" and off limits for new reservoirs or diversions. Another casualty would be the peripheral canal, which eventually drew strong environmental opposition.

The wave of new environmental legislation began in the mid-1960s, with a succession of federal laws regarding water quality and species protection—the National Wilderness Preservation Act (1964), the Federal Endangered Species Preservation Act (1966, a precursor to the 1973 Endangered Species Act), the National Wild and Scenic Rivers Act (1968), the National Environmental Policy Act (1969), the Clean Water Act (1972), and the Safe Drinking Water Act (1974). California's legislature was equally active in the environmental arena, passing comparable bills at the state level.

As species protection became an explicit goal in the Delta, alongside the maintenance of fresh water for human uses, perceptions of the effects of water diversions and the nature of water quality problems began to change. In 1971, the State Water Resources Control Board (SWRCB) adopted Water Rights Decision 1379, establishing water quality standards for the CVP and the SWP that included new outflow requirements for the San Francisco Bay–Delta Estuary and a comprehensive monitoring program to follow changes in environmental conditions. This decision, stayed by court order in response to lawsuits filed by San Joaquin Valley irrigation districts, marked the beginning of a series of legal and regulatory battles over Delta water quality standards for the environment.[11]

DEFEAT OF THE PERIPHERAL CANAL

During the 1970s, the peripheral canal plan was also subject to increased environmental scrutiny. Although the canal was initially promoted as having environmental benefits in addition to the primary benefit of controlling the salinity of Delta water exports, these benefits were not spelled out in any detail in the reports of the 1960s. Subsequent reports were more mixed. Controversy around the plan began to build, generating considerable debate, including lawsuits, over several years.[12] In the end, the canal was beaten in the court of public opinion. By the time it was put to a referendum in 1982, an alliance of environmentalists and northern water interests, with backing from some Tulare Basin farmers who feared high water costs (Arax and Wartzman 2003), successfully argued that the canal would be bad for the environment and Northern California water rights. Large majorities of Northern California voters rejected the perceived water grab by Southern California (Figure 2.4).

DROUGHT INTENSIFIES CONFLICT

In 1987, California entered a multiyear drought that severely reduced available flows from the Delta's two main watersheds. As the drought wore on, it provoked conflict over the amount of water reserved for environmental flows. Initially, CVP and SWP exports were not cut, and both environmentalists and fisheries agencies raised concerns over the consequences for important fish species that depended on the Delta. In 1989, the Sacramento River winter-run Chinook salmon was listed as threatened under the federal Endangered Species Act and as endangered under its state counterpart, and DWR and USBR agreed to build salinity control gates in Suisun Marsh and make other efforts to preserve the habitat in the marsh.

With the drought still in full force, water exports to some San Joaquin Valley farmers were reduced in 1991 to maintain minimum environmental flows. The following year, water users were dealt several legal and legislative blows. The courts upheld that an irrigation district must cease pumping during peak migration times for endangered Chinook salmon and that the CVP must release flows sufficient to protect downstream fisheries. Congress then passed the Central Valley Project Improvement Act (CVPIA), which included a requirement that the CVP commit 800,000 acre-feet of water per year (roughly 10 percent of total deliveries) to support fish and wildlife.

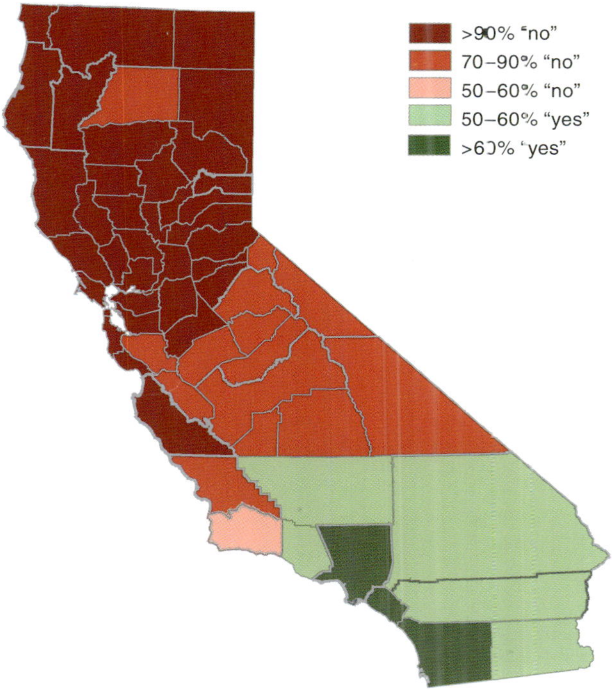

FIGURE 2.4 County voting patterns on Proposition 9 (Peripheral Canal), June 1982. Source: California Secretary of State 1982.

By 1993, a crisis was erupting. The delta smelt was listed as a threatened species, and other listings began to follow (Table 2.2). The federal EPA (Environmental Protection Agency) threatened to impose stricter water quality standards for the estuary that would severely curtail water exports. Under the threat of a regulatory hammer, water users agreed to work with environmental interests to forge a new plan for the Delta that would comprehensively address both water user and environmental concerns. In December 1994, the signing of the Bay-Delta Accord marked the beginning of the CALFED era.

THE CALFED ERA: TESTING THE LIMITS OF CONSENSUS

CALFED sought to involve the full array of relevant federal and state agencies, together with local and statewide stakeholders, to form a new plan for the Bay-Delta. The CALFED process continued in earnest for roughly

TABLE 2.2

Status of Fish Species in the Sacramento–San Joaquin Delta Watersheds

Species	*Year*	*Status*
Sacramento River winter-run Chinook salmon	1989	Endangered (CESA) Threatened (ESA)
Delta smelt	1993	Threatened (ESA and CESA)
Sacramento River winter-run Chinook salmon	1994	Reclassified as endangered (ESA)
Sacramento splittail	1995	Species of concern (CESA)[a]
Longfin smelt	1995	Species of concern (CESA)
Sacramento perch	1995	Species of concern (CESA)
River lamprey	1995	Species of concern (CESA)
Central Valley steelhead trout	1998	Threatened (ESA)
Central Valley spring-run Chinook salmon	1999	Threatened (ESA)
Sacramento River drainage spring-run Chinook salmon	1999	Threatened (CESA)
Central Valley fall-run and late-fall-run Chinook salmon	2004	Species of concern (ESA)
Southern green sturgeon	2006	Threatened (ESA)
Longfin smelt	2009	Reclassified as Threatened (CESA)
Delta smelt	2009	Reclassified as Endangered (CESA)

SOURCE: Department of Fish and Game 2006a.

NOTE: ESA and CESA refer to the federal and California Endangered Species Acts, respectively.

[a]The Sacramento splittail was listed as threatened under the ESA in 1999 but was removed from the list in 2003.

a decade, funded primarily by state bond monies with some federal contributions.

One of CALFED's early efforts was to review and compare strategic alternatives for the Delta. Over 20 diverse conceptual alternatives were initially reviewed and briefly discussed, but little formal analysis was published (CALFED 1996). The CALFED Record of Decision (ROD) was signed in mid-2000 by all agencies with authority over Delta operations, and it advocated the continuation of the through-Delta strategy for water exports. All four of CALFED's main goals (water supply reliability, water quality, ecosystem restoration, and levees) were based on this strategy and were not to be revisited until 2007. The maxim that "everyone would get better together" tied all fates to this single approach.

CALFED proved to be a fragile truce. By the tenth anniversary of the Bay-Delta Accord, stakeholder frustrations were widespread. Water export users were frustrated with slow movement to augment water supplies, which in some cases meant restoring supplies that had been reduced to support the environment. In-Delta users were discouraged by the limited progress on dealing with Delta salinity and water quality. Environmental interests remained concerned that water export goals were taking precedence over ecosystem protection—a concern that turned into alarm when the news broke about precipitous drops in the delta smelt and other pelagic fish species. And Delta landowners and farmers were frustrated over limited funds for levee improvements and maintenance, which had previously received some state funding but were not a priority for CALFED funds.

Arguably, CALFED was not designed to deal with some of the problems that have recently emerged. New research on the long-term risks associated with Delta levees, the significant levee breach on Jones Tract in the summer of 2004, and the devastating effects of levee breaches in New Orleans all made the levee issue more urgent than it had been in the years leading up to the CALFED ROD. Similarly, CALFED's initial ecosystem focus was on restoring salmon runs, in part because delta smelt and other pelagic organisms were less understood. The recent severe declines in these fish populations caught most experts by surprise.

CALFED was also founded on the implicit assumption that the Delta would not face the urbanization pressures that have become apparent over the past few years. This assumption may have been justified in the early to mid-1990s, particularly in light of the passage of the Delta Protection Act of 1992, which reserved most Delta lowlands for agricultural and environmental uses. Beginning in the late 1990s, a housing boom swept the Central Valley, and a number of large projects are slated for development in lowland areas that are exempt from the act's restrictions (Eisenstein et al. 2007). In addition, recent concerns about urban flood risks behind agricultural levees, state liability for failure of project levees (following the 2003 *Paterno* decision), and the long-term environmental effects of urbanizing Delta islands have raised urbanization as a serious long-term issue for Delta management.[13] Although housing growth has virtually ground to a halt since the onset of the recession in 2007, these regulatory gaps will once again pose challenges when market conditions improve.

But CALFED also suffered from some fundamental design flaws, particularly with regard to financing. CALFED parties agreed to a principle of "beneficiary pays," but in practice, the implications for user contributions were never fleshed out. The program was launched at the height of the dot-com boom, when the state enjoyed windfall surplus revenues, and it relied on unrealistic expectations of massive state and federal taxpayer funds. Serious, long-term funding proposals were never developed. This lack did not matter so much in the first years after the signing of the ROD, because $1.5 billion in state bond funds was earmarked for the program (de Alth and Rueben 2005). But by 2005, when most bond funds had run out, legislative frustration over the lack of a realistic plan for beneficiary contributions spelled the end of most CALFED activities.

CALFED did achieve some notable successes. Major improvements were achieved in interagency coordination. Considerable progress was made in ecosystem restoration in several watersheds upstream of the Delta. Water transfers have become largely accepted statewide, with success during the 1987 to 1992 drought followed by a successful Environmental Water Account (Hanak 2003). Improvements in water conservation efforts have continued, and funding for research has brought more data and some new thinking to Delta ecological problems. Ultimately, however, the program suffered from a failure of political processes to come to long-term agreement without continued massive taxpayer subsidies. In light of the new problems facing the Delta, it now appears that the CALFED premise that everyone can get better together may be unrealistic.

NEW INITIATIVES AND NEW TROUBLES IN THE DELTA

Since 2007, several policy initiatives have been under way to develop sustainable solutions to Delta problems. The governor's "Delta Vision" initiative, led by an independent Blue Ribbon Task Force reporting to a cabinet-level committee, released several reports outlining a strategic vision and long-term plan (Isenberg et al. 2008a, 2008b, Natural Resources Agency, 2009). Separately, water export agencies, fisheries agencies, and many environmental and local interest groups are working to develop a Bay Delta Conservation Plan (BDCP)—a habitat conservation plan for improving the management of protected aquatic species that will regulate

Delta exports. Meanwhile, the legislature has been exploring options for improving Delta ecosystem and water management through a special session on water, water bond discussions, and several proposed bills.[14]

Each of these efforts has highlighted the importance of balancing water supply and ecosystem needs. Each effort is exploring the possibility of building a peripheral canal, most likely to be managed as a "dual" conveyance system with some continued Delta export pumping. To support the decision process regarding Delta conveyance, the governor's office recently instructed the Department of Water Resources to analyze these and other alternatives as part of an environmental impact review.

The sense of urgency has been heightened by deteriorating conditions for key Delta fish species. Since December 2007, the export pumps in the southern Delta have been operating at reduced levels under an order from federal Judge Oliver Wanger because the Central Valley Project and the State Water Project were found to be in violation of the federal Endangered Species Act regarding delta smelt.[15] Spring 2007 population counts for this species, which have been in sharp decline since 2004, registered another precipitous drop, raising fears that the smelt are bordering on extinction. Export users estimate that the court-imposed interim rules will reduce water exports by roughly 22 to 30 percent (Department of Water Resources 2007b). As outlined in the new "biological opinion" for the delta smelt, issued in December 2008, the longer term solution may require further cutbacks (U.S. Fish and Wildlife Service 2008).[16] Additional cutbacks may be ordered as the result of a second decision by Judge Wanger (April 2008) to protect the listed winter- and spring-run Chinook salmon, both in decline.[17]

Other key species also are in trouble. After fall 2007 surveys recorded the lowest numbers on record for longfin smelt—another pelagic (open-water) species that lives in the Delta—it was listed as a threatended species under the California Endangered Species Act (CESA) in March 2009. This action likely implies additional regulations at the pumps. Since 2006, there has also been a rapid decline of fall-run Chinook salmon, forcing closure of the ocean commercial and sport fisheries in California and most of Oregon for 2008 and 2009.

In the midst of what water export users began calling a "regulatory drought," rainfall conditions also deteriorated, leaving reservoirs at some of their lowest levels in history, and forcing cropping cutbacks and urban rationing in many areas dependent on Delta exports. Thus, conditions are now similar, in many ways, to the dire circumstances facing water users during the early 1990s drought. That crisis pushed parties to the negotiating table, but the resulting CALFED process proved too weak to provide a durable solution. The question now before California's leadership and stakeholders is whether they can build on the lessons of the past to institute lasting reform of Delta management.

THE LESSONS OF DELTA HISTORY

The Delta's short history of European settlement has seen major changes in the form, use, and settlement of land in the Delta. Before European settlement, the Delta was a massive tidal marsh, with significant seasonal variations in flow and salinity, as well as large interannual variations caused by floods and droughts. This era was followed by a period of land reclamation for agriculture, which, for better or worse, created much of the Delta's current landscape. Marsh reclamation reduced tidal flows, but upstream diversions in the Sacramento Valley increased salinity intrusion into the central Delta during dry seasons of dry years, processes clearly understood in the 1930s.

The prospect of major water exports from the Delta made salinity intrusion a primary concern for all water users within the Delta. Various strategies, including saltwater barriers, were considered early on. By the 1930s, a hydraulic barrier, consisting of Delta outflows from upstream reservoirs, was selected as the primary means of salinity control for agricultural and urban water users. Using this approach, both in-Delta users and water export users could agree on a need to keep the Delta always fresh.

The notion of an always-fresh Delta supported by persistent net Delta outflows has endured for over 70 years, but it is not aging well. This management strategy retains support from in-Delta users, but water export users have come to see increasing risks from this approach, for reasons described in Chapter 3. In Chapter 6, we will examine changes in our understanding of the Delta ecosystem, which also cause us to doubt the wisdom of continuing with this strategy. Because of the history of profound and widespread change in the Delta, it is long past the point where the Delta can be "restored" to past conditions, whether it be the pre-European Delta or the bucolic agricultural Delta. No matter which course is chosen, the Delta of the near future will be very different from past Deltas.

Delta history provides insight into the processes by which Californians have sought solutions to collective problems in this pivotal region. And as this history suggests, these processes have rarely been simple or smooth. At several points over the last century, strenuous efforts have been made to provide solutions to the Delta's problems, and these solutions have been followed by major investments in the chosen strategy. From the 1890s to the 1910s, the Debris Commission worked on Central Valley flood control. Later, state and federal efforts developed the 1930 State Water Plan and executed the Central Valley Project; investigations in the 1950s led to the development of the State Water Project. In more recent times, as environmental concerns have become central in Delta policy considerations, the search for solutions appears more constrained. Thus, CALFED worked under the premise that the Delta's basic configuration should remain unchanged and that environmental goals could be satisfied simultaneously with those of export users and in-Delta interests. Given the crisis now looming in the Delta, it is once again time for California to launch a serious search for solutions for a new Delta.

3

MANAGING THE INEVITABLE

> Assuming no public aid, it is conceivable that the exhausting peat will cause land to subside to the point where drainage and levee maintenance costs will make continued operations impracticable.
>
> JOHN THOMPSON (1957), *The Settlement Geography of the Sacramento–San Joaquin Delta, California*

The Sacramento–San Joaquin Delta is significantly changed from its historic condition. Before the arrival of Europeans, the Delta was one of California's most dynamic landscapes. Lying at the confluence of the Sacramento and San Joaquin rivers and their floodplains at the head of the San Francisco Estuary, with its extensive marshes and tidal channels, "equilibrium" in the Delta consisted of constant change, involving self-adjustment to daily tides, annual floods and droughts, shifts in climate, and rise in sea level. As noted in Chapter 2, the policies and practices of the past 150 years have overtly or inadvertently reduced the Delta's dynamic character. Draining of marshes and floodplains through construction and maintenance of 1,100 miles of levees sought to freeze the landscape in place, ending the historical connection between landscape-shaping water flows and the landscape itself. For decades, the Delta's waters and lands have been managed in this artificial way, by fixing levees when they fail and changing releases from upstream reservoirs and the timing of water exports to keep the Delta's interior fresh.

Natural systems that have been heavily influenced by human activities—of which the Delta is a prime example—are vulnerable to significant change from two distinct sources (Berkes and Folke 1998). First, external factors (such as climate change) can dramatically alter conditions. Second, human activities can interact with natural processes to create additional pressures. As these systems lose their resiliency, they become vulnerable to dramatic and potentially abrupt shifts in physical and biological conditions (Liu et al. 2007). These changes often have thresholds or tipping points where change is both fundamental and irreversible, defining a new regime or state (Gunderson and Holling 2001).

The Delta is at a tipping point. The transition between the tidal freshwater marsh of the early nineteenth century to the subsided island, tidal channel network of the twentieth century was a dramatic and rapid change for this region. Today's Delta is unstable and poised for another major change. Subsidence, sea-level rise, changing inflows, and earthquakes, along with the escalating costs of resisting these processes, are shifting the Delta toward a markedly different state from that of the nineteenth and twentieth centuries. As outlined below and in Suddeth et al. (2008) and Suddeth (2009), these physical and economic drivers of change will transition the Delta during this century to a system with significant amounts of open water as islands flood permanently. This shift in conditions, and the uncertain responses of the Delta's many aquatic and terrestrial species, poses an unprecedented management and policy challenge for California.

DRIVERS OF CHANGE

The landscape of the Delta is defined by its levees. Failures of these levees, whether planned or unplanned, will transform the Delta in the future. The risk of levee failure in the Delta today is, by all standards, quite high. The causes of these failures are varied, including overtopping, erosion, through-seepage (which erodes the levee interior), under-seepage (which erodes the foundation), slumping of embankments, compaction, and foundation failures during earthquakes. Although current risk factors are high today, four physical factors are directly or indirectly causing levee failure to become increasing likely in the future. These physical drivers of change are subsidence, changing inflows, sea-level rise, and earthquakes.

SUBSIDENCE

The most dramatic landscape change in the Delta is subsidence: the lowering of the land surface throughout most of the Delta as a result of the loss of organic-rich soils during farming (Drexler et al. 2007). Soil loss is principally caused by microbial oxidation of the peat when water tables are lowered below the crop root zone, with lesser amounts of subsidence caused by compaction, erosion, and historic peat burning. The amount of subsidence in the Delta is extraordinary. In an earlier study, Mount and Twiss (2005) demonstrated that the volume of soil lost in the islands exceeds 3.4 billion cubic yards (enough to fill 8,000 Rose Bowls), with average elevations of islands commonly more than 15 feet below sea level (see Figure 2.3). Moreover, Mount and Twiss (2005) modeled future subsidence of the Delta and projected that in the next 50 years a potential additional 650 million cubic yards of soil will be lost. For perspective, if the cost to deliver a cubic yard of soil to a Delta island is roughly \$25, the cost to replace this soil would be roughly \$325 million annually.

Subsidence increases the hydraulic forces that cause levee failure and island flooding by increasing the difference in elevation between the island interior and the water surface in adjacent channels (Figure 3.1). Subsidence also increases the consequences of major levee failures, causing more saline water to be pulled into the Delta from the San Francisco Bay to extend the interruption of water supply exports. Subsidence also increases levee repair costs by expanding the size and depth of levee breaches and increasing the amount of water that must be pumped from flooded islands, as well as increasing island drainage pumping costs if levees do not fail.

CHANGING INFLOWS

Broad scientific evidence indicates that California's hydrology is changing. A recent synthesis by Barnett et al. (2008) of the hydrology of the western United States confirms a trend noted by several authors in the last decade. Most notably, the mix of rain and snow in mountainous regions has been shifting for much of the past 50 years, leading to more winter runoff, with an earlier spring snowmelt. The intensity of winter floods also has increased during this period, consistent with this shift of runoff to the winter (Stewart et al. 2004). Climate projections for California are in broad agreement that current trends toward earlier runoff will continue (Cayan et al. 2008b). This shift in timing may be accompanied by increases in the interannual variation and magnitude of floods (Maurer 2007). A Mediterranean climate that is naturally variable appears to be increasing in variability.

As Florsheim and Dettinger (2007) have demonstrated, even with significant improvements in levees, the frequency of levee failures in the Central

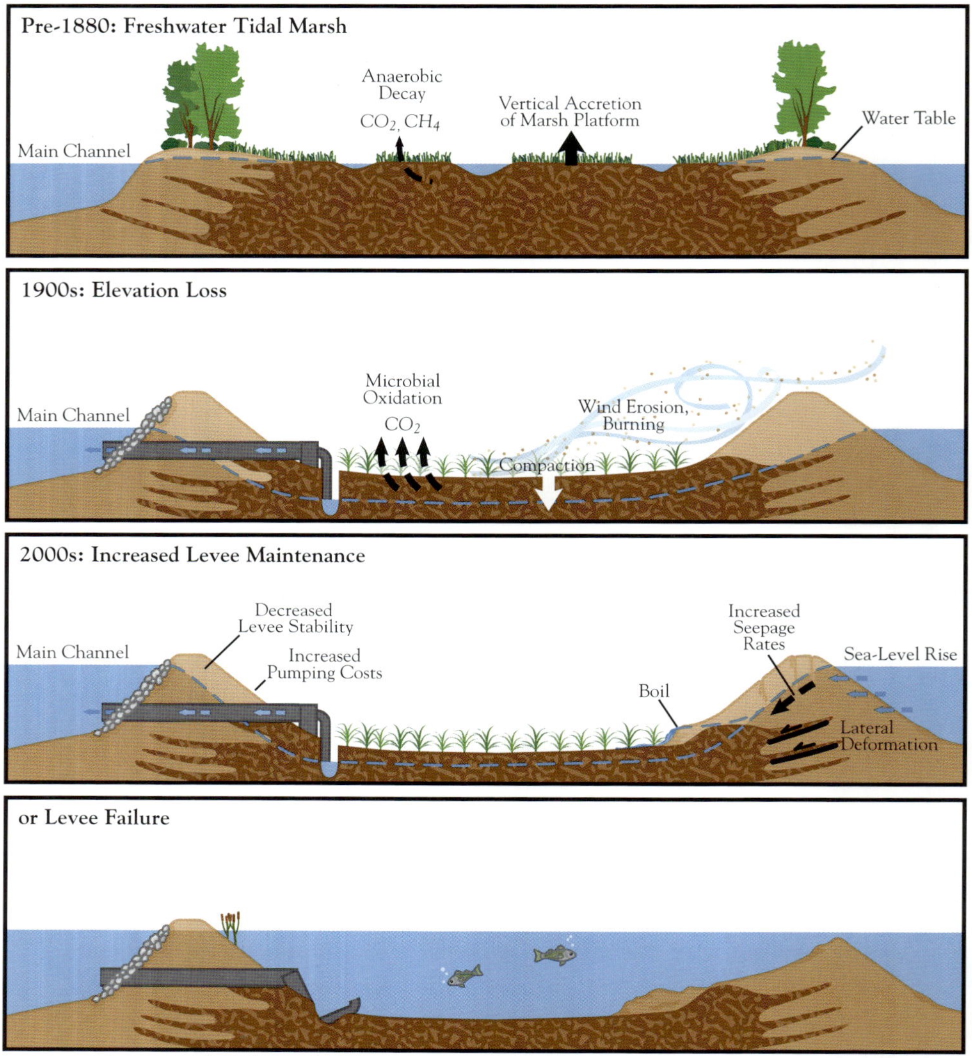

FIGURE 3.1 Historic and future trajectory of island subsidence in the Delta. Source: Mount and Twiss 2005.

Valley, and in the Delta specifically, are tied directly to the frequency of large storms. Flows both within and outside the Delta remain confined to a narrow network of channels with limited capacity. By confining flows, the levees increase the water surface elevation, increasing the hydraulic gradients within the very levees that confine the flow. Thus, future increases in intensity of winter inflows to the Delta, coupled with the poorly built, narrow levee network, will increase the frequency of levee failures.

SEA-LEVEL RISE

The Delta was constructed naturally by sea-level rise. Over the past 6,000 years, a slow rise in sea level "drowned" the lower reaches of the Sacramento and San Joaquin rivers, forming the current estuary and allowing accumulation of vast quantities of organic material in the vicinity of the Delta (Atwater et al. 1979). The past century has shown a well-documented rise in sea level (Church and White 2006) that has directly affected the San Francisco Estuary. Sea level is tied to all aspects of the estuary, especially its elevations and tidal hydrodynamics (Krone 1979). The future rates of sea-level rise are the subject of considerable scientific debate. At the request of the CALFED Lead Scientist, the CALFED Independent Science Board reviewed this issue and made recommendations for projections of future sea-level rise (Mount 2007). They recommended that empirical models be used to project sea-level rise until physical models are improved. Following the work of Rahmstorf (2007), they recommended that planning for the Delta should accommodate a midrange value of roughly 28 to 39 inches in sea-level rise by 2100, with approximately one-third of that rise occurring by 2050 (Figure 3.2).

One foot or more of sea-level rise will be a dramatic change for the Delta. Most Delta levees have only one foot of freeboard over the estimated 100-year water surface elevation, making them vulnerable to overtopping and increased seepage. Cayan et al. (2008a) note that the effect of sea-level rise will go beyond a rise in mean water surface elevations with associated increased pressures on levees. Their modeling has shown that sea-level rise will increase the frequency, magnitude and duration of extreme high-water events. These events are formed by the co-occurrence of high tidal elevations, El Niño–like disturbances, low pressure systems, and high inflows. This increase in the frequency and length of time levees are stressed by high water elevations will significantly raise the likelihood of failures.

EARTHQUAKES

The Delta lies within a region known worldwide for its high risk from earthquakes. This issue has been of concern to the Department of Water

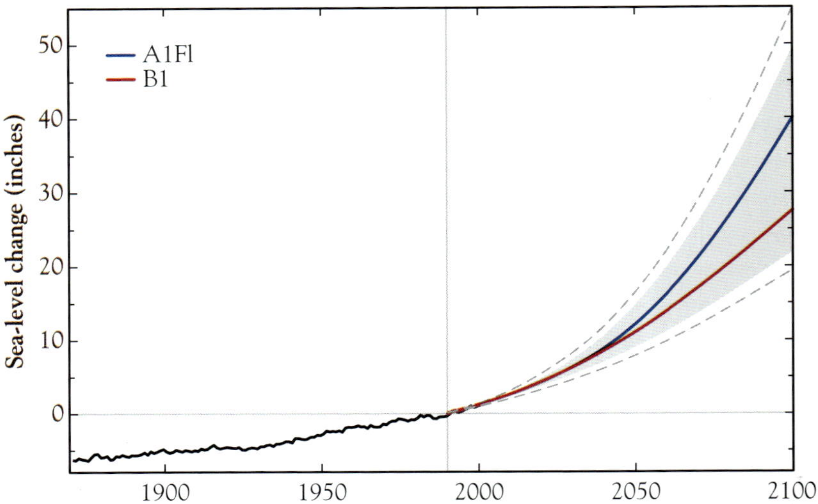

FIGURE 3.2 Historical and projected range of sea-level rise, based on empirical relationship between climate warming and sea surface elevations. A1Fl and B1 are greenhouse gas emission-related warming scenarios from the Intergovernmental Panel on Climate Change. Source: Modified from Rahmstorf 2007.

Resources for more than 30 years because of the potential for multiple levee failures during an earthquake and a shutdown of the State Water Project. The most recently completed review of the risk of levee failure from earthquakes (Torres et al. 2000), which used limited geotechnical information regarding levees and their foundations, showed that ground accelerations from earthquakes with 100-year recurrence intervals (annual probability of 1 percent) are sufficient to cause multi-island flooding. The failure risk is highest in the western Delta, where levees close to possible earthquake faults are poorly constructed and on weak foundations.

A more comprehensive assessment of Delta earthquake risks was completed as part of the Department of Water Resources Delta Risk Management Strategy (DRMS) program. The DRMS assessments show exceptionally high probabilities of failure from earthquakes, with indications that ground accelerations from earthquakes with 20-year recurrence intervals (5 percent annual probability) can generate multi-island failures (URS Corporation and Jack R. Benjamin and Associates 2009). This is consistent with estimations of a 62 percent probability of a large earthquake in the Bay

Area within the next 30 years (U.S. Geological Survey [USGS] 2003). The probability of a major earthquake occurring in the region also increases with time as stress builds on Bay Area fault systems.

MANAGING OR RESISTING CHANGE

Together, the forces acting on the Delta's levees—subsidence, changing inflows, sea-level rise, and earthquakes—are attempting to shift the Delta from its current configuration as a network of levees protecting subsided islands toward an expanse of open water fringed by tidal marsh. The timing, magnitude, and location of change in the Delta will depend in part on society's investments to resist or accommodate these changes, whether through levee improvements or repairs following failure.

Contrary to popular belief outside of Delta circles, the network of levees is not a unified system maintained or overseen by a single entity. Approximately one-third of the 1,100 miles of levee in the Delta are "project" levees. These levees are included within the federally authorized Sacramento-San Joaquin Flood Control Project (see Figure 2.1) and are maintained by local reclamation districts overseen by the state of California. The remaining two-thirds of the Delta's levees, including most in the heart of the Delta, are "nonproject"; that is, owned and managed by local reclamation districts on behalf of island landowners.

There is also a common misconception about the level of investment by state and federal agencies in the Delta levee system. Although there have been substantial state and federal funds for project levees, nonproject levees have been constructed principally with private and local funds. The most important state investments in Delta levees today are addressed by the Delta Levee Subventions Program. This program began after a series of levee failures highlighted the poor condition of the nonproject levees in the Delta and the risk they posed to the state water supply system. The Subventions Program helps defray some costs of maintenance, repair and upgrades of nonproject levees. This modest subvention is typically less than $10 million per year[1] and has leveraged substantial investments in the levees by local reclamation districts. This program has significantly reduced (but not eliminated) levee failures.

Large costs to the state and federal government can occur when levees fail. The damages and economic costs from the recent Jones Tract failure

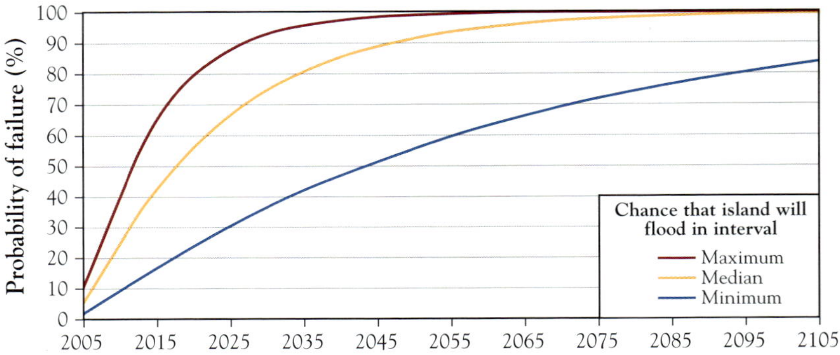

FIGURE 3.3 Cumulative likelihood of island flooding over time for 36 significant Delta islands. Chart plots the cumulative probability of failure from earthquakes or floods sometime during the period up to years shown.Source: Author's calculations, using data from URS Corporation and J.R. Benjamin and Associates 2007b. Chart plots the cumulative probability of failure from earthquakes or floods sometime during the period up to years shown.

in June 2004 exceeded $90 million, with the State of California alone responsible for more than $30 million in flood fight, levee repair, and island pump-out costs.

Even large state investments in levees are unlikely to significantly alter the trajectory of change in the Delta. To illustrate the magnitude of this problem, we calculated the annual probability of island flooding from floods and earthquakes of 36 subsided islands that make up two-thirds of the Delta (summary in Suddeth et al. 2008 and Suddeth 2009). These annual probabilities are based on a DRMS report and account only for today's risk, not increasing future risk. Thus, this analysis represents a business-as-usual approach, assuming continued investments in island maintenance and repairs. Although the probability of an island flooding in any single year may be low (less than 5 percent for a typical Delta island), the *cumulative* probability of an island flooding some time in the next 50 years is much higher. Figure 3.3 plots the likelihood of island failure, based on present risk factors: the further into the future, the higher the accumulated likelihood of failure. This estimation shows that Delta islands facing the median risk of failure have a 95 percent chance of failure from floods or earthquakes some time in the next 50 years. Half of the islands have a higher probability of failure over this time period. The western islands are most at risk, with a 98 percent chance of failure some time before mid-century. Incorporating future increases in risk from subsidence, sea-level rise, inflows, and earthquakes would increase this already high probability of failure.

The societal response to the extraordinarily high likelihood that many islands will fail will presumably depend on the costs of mitigating risk. To give a sense of the sums involved, the costs and benefits of upgrading and repairing the 34 core Delta agricultural islands were evaluated (Suddeth et al. 2008).[2] The costs of upgrading the islands to federally mandated standards for Delta agricultural levees are substantial. This standard, known as PL (public law) 84-99, raises levees approximately six inches over basic state Hazard Mitigation Plan standards and reduces their interior side slope to increase stability and resist seepage erosion. Upgrading to PL 84-99 standards was a goal for the CALFED levee program, but few funds were actually allocated to the effort. In today's dollars, the cost of this upgrade is estimated to exceed $1.4 billion. This sum could be an underestimate, because it presumes that the project levees already meet this standard, leaving only nonproject levees to be upgraded. Moreover, as the Jones Tract example illustrates, levee repairs for an individual failure can run into the tens of millions of dollars.

To help assess how best to prioritize public spending on Delta levee upgrades and repairs, we developed a simplified Levee Decision Analysis Model. Using annual failure probabilities for today's conditions and values for land and other assets contained in DRMS documents and other sources, we estimated net expected costs for: (1) no upgrading of current levees, (2) upgrading levees to PL 84-99 standards, (3) adding an additional foot in height above PL 84-99 standards to protect levees from sea-level rise (at a 20 percent premium over PL 84-99 costs), and (4) repairing versus not repairing island levees once they fail. The decision to make levee upgrades or repairs weighs the costs of these investments against the economic value of the assets on the island. This approach is similar to analyses developed by the Dutch for flood management (van Dantzig 1956; Voortman et al. 2002).

Where the net expected costs of levee upgrades and repairs exceed the expected economic benefits of an island, additional upgrade and repair investments are unlikely to be justifiable from an economic perspective.

Local reclamation districts and the individual landowners on these islands would presumably not be restricted from making such investments in their property, although this would likely be an unsound business investment. This analysis assumes that the state would continue modest investments in levee upkeep through the subventions program, regardless of the upgrade or repair status of any levee.

Although these calculations are preliminary and represent one of many potential approaches to prioritizing levee investments, they are sufficient to demonstrate several key points. First, based solely on net expected cost, it is most cost-effective for *all* islands analyzed here to forgo investment in major levee *upgrades*. This surprising result arises because levee improvement costs are high; upgrade costs often exceed the current land value of the island. Even if land and asset values were increased significantly, upgrades remain uneconomical. Also, even rather expensive upgrades provide only marginal improvements in levee reliability and do not substantially reduce the frequency of levee failures. According to Delta engineers, upgrading to PL 84-99 standards reduces failure probabilities from high-water events by only approximately 10 percent and does little to improve performance during earthquakes. Implausibly large increases in levee reliability would be needed to make Delta levee upgrades economically worthwhile.

Second, it is not cost-effective to *repair* many Delta islands after levee failures. In Figure 3.4, the decision on levee repairs is weighed against the economic value of land (mainly farmland) and other assets (houses and other buildings plus roads and other infrastructure).[3] Figure 3.4 also highlights the five western islands that may be critical for maintaining through-Delta exports, as described in Chapter 5. Islands are classified as "do not repair" when the costs of *not* repairing a levee breach (including loss of the assets and mitigation of potential effects on levees on neighboring islands) are lower than the costs of repair by a margin of at least 100 percent. For islands classified as "repair," the inverse is true. For islands classified "indeterminate," the difference in cost between repairing and not repairing the island was less than twice the cost of the cheaper strategy. Roughly half the islands fall into the "do not repair" category following failure. For many of these islands, repair costs are high relative to overall island value.[4] When the costs of long-term infrastructure maintenance and upgrades are factored in (for instance, replacement of bridges, road widening), it may also prove less expensive to relocate infrastructure rather than paying for successive repairs on some islands.[5]

This analysis excludes islands with substantial urban settlements, such as Bethel Island (#2) and Hotchkiss Island (#23) (shown in green in Figure 3.4). Such locations would likely merit upgrading to higher urban standards to protect property and improve public safety. Urbanization of Delta lands raises important issues with respect to flood management and availability of upland habitat, as discussed in Chapter 8.

For the included islands, the analysis does not account for their importance for purposes other than local economic activity (primarily farming) and infrastructure (notably road and rail). We did not consider the value of specific islands for terrestrial or aquatic habitat, such as sandhill crane habitat on Staten Island (#55) or social and cultural values, such as legacy towns like Isleton and Walnut Grove. Inclusion of such values—which are difficult to quantify—could raise the cost of protecting some islands. Nor did we include the increasing drainage and pumping costs for Delta islands likely to accompany continued subsidence and sea-level rise, which will probably lower Delta land values over time.

Finally, and perhaps most significantly for Delta water management debates, we did not explicitly consider the value of Delta islands for maintaining water quality for export facilities and in-Delta uses. However, as discussed further in Chapter 5, water quality modeling efforts suggest that only the westernmost islands could be critical to maintaining water quality for through-Delta exports (see shaded islands in Figure 3.4). In contrast, islands in the central, eastern, and southern Delta might remain flooded following levee failure without significant effects on the quality of water exported from the Delta.[6] This finding is contrary to conventional wisdom, which assumes that most islands of the western, central, and southern Delta are important for maintaining water exports.

The DRMS study estimates that upgrading the western islands to be either repairable following a major earthquake or resistant to earthquakes would cost from $3.6 billion to $5.2 billion, respectively. As discussed in Chapter 5, the benefits of these upgrades would be limited in time, because sea-level rise will eventually degrade Delta water quality or require greater net outflows even if the western islands remain intact.

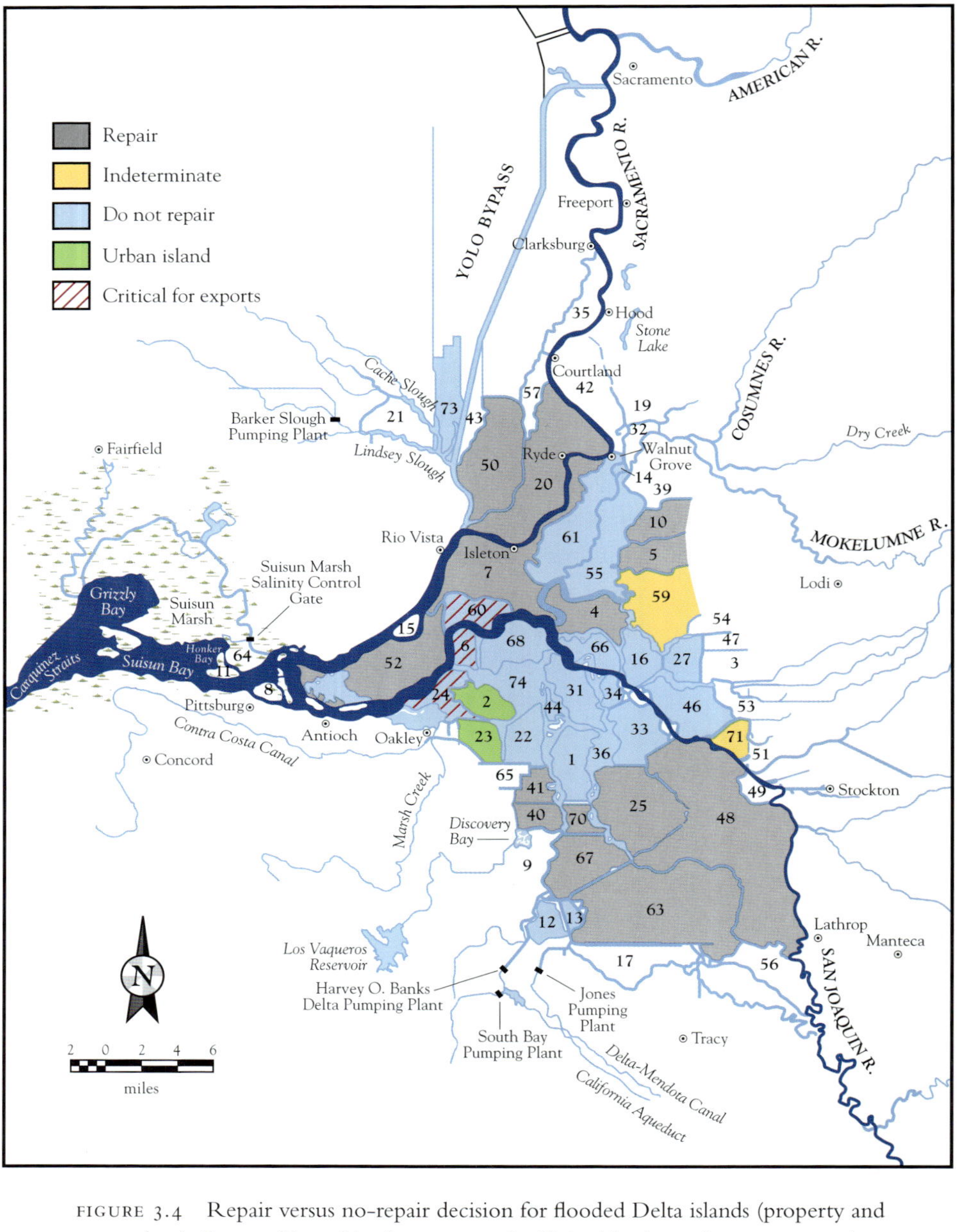

FIGURE 3.4 Repair versus no-repair decision for flooded Delta islands (property and asset values). For a guide to island names, see the Delta island map key on page xxi. Source: Suddeth 2009 and Suddeth et al. 2008.

A FUTURE DIFFERENT FROM THE PAST

The current Delta, with its elaborate network of channels, subsided islands, and managed inflows and exports, is an unstable system highly vulnerable to perturbation. Geologically, this vulnerability is reflected in the high probability of island flooding. Biologically, this vulnerability is reflected in the recent decline in many native Delta species. If current risks hold, many Delta islands are likely to flood within the next several decades. Unmanageable processes external to the Delta, such as earthquakes, climate change–influenced sea-level rise, and increased flood flows, are being exacerbated by human activities, such as land uses that increase subsidence. All these drivers point to increased risks of Delta island flooding.

The resiliency of the Delta in its current state depends on its ability to self-adjust and the ability and willingness of Californians to maintain all or parts of the status quo. Unlike natural estuarine or floodplain systems, the Delta's artificial landscape cannot self-adjust to rising sea levels, earthquakes, land subsidence, and higher inflow extremes. As others have also found (Logan 1989, 1990a, 1990b), the cost of maintaining current levees outstrips the value of many Delta islands, depletes all available bond funds, and, in all likelihood, dampens the willingness of the public to continue subsidies as failures and costs escalate. It is more cost-effective to invest selectively in Delta levees to protect high-value lands and critical infrastructure. Furthermore, depending on the final conveyance choice, it is cost-effective to invest in islands that support export water of acceptable quality, and to let lower-value islands without compelling state interests eventually flood and return to aquatic habitat.

The combination of natural forces and artificial conditions created by human activities points toward an inescapable conclusion: The Delta is undergoing a fundamental transformation—one that could accelerate with catastrophic levee failures. The new Delta will significantly differ from the marshy Delta of the early nineteenth century or the agricultural Delta of the twentieth century. The third-generation Delta of the twenty-first century will likely have large tracts of open, deep (more than 20 feet) water, particularly in the western and central Delta. Islands crucial to state infrastructure, water supply, or ecosystem goals will presumably be prioritized for public investments in maintaining and upgrading levees or modifying or relocating infrastructure. This fundamental and inevitable change will have significant ramifications for all aspects of Delta land and water use.

4

DELTA WATER EXPORTS AND STRATEGIES

> It is probable that serious objections would be raised to any structure which might have the effect of seriously disturbing or eliminating commercial and recreational fishing in the upper bay and rivers.
>
> RAYMOND MATTHEW (1931), *Economic Aspects of a Salt Water Barrier Below the Confluence of Sacramento and San Joaquin Rivers*

Changes in the Delta are inevitable, given the unstoppable processes of sea-level rise, land subsidence, earthquakes, and a warming climate bringing larger floods. As discussed in Chapter 3, these changes pose grave questions about future land uses in many parts of the Delta. Anticipating these changes is also critical for managing California's water supplies, given the Delta's central role in moving water from Northern California watersheds to farmlands and cities south and west of the Delta. Recent water exports from the Delta have ranged from 5 to 6 million acre-feet (maf) per year, supplying much of the water used in the Bay Area, the southern Central Valley, and Southern California. On average, these exports account for about 15 percent of natural flows into the Delta watershed and 25 percent of Delta inflows after reductions from upstream diversions.

Decisions made about how best to manage water exports from the Delta will determine much of the future of California's water system and will be central in defining opportunities and alternatives for managing the Delta's ecosystem.

STATE AND REGIONAL USE OF DELTA WATER SUPPLIES

Table 4.1 presents estimates of the consumptive uses of water (water that is either consumed or evaporated and is unavailable for potential reuse) in

TABLE 4.1

Estimated Average Consumptive Uses of Delta and Delta Tributary Waters (taf/year), 1995–2005

Demand Area	*Agriculture*	*Urban*	*Environment*[a]	*Total*
Natural flow	—	—	40,293	40,293
Net Delta outflow	—	—	22,553	22,553
Total diversions	14,090	3,235	415	17,740
Upstream diversions	9,540	1,712	138	11,390
Delta diversions	4,550	1,523	277	6,350
In-Delta diversions	769	0	0	769
Upstream diversions	0	0	0	0
Delta diversions	769	—	—	769
North of Delta diversions	6,000	562	138	6,700
Upstream diversions	6,000	520	138	6,658
Delta diversions	0	42	0	42
South of Delta diversions	7,321	1,699	277	9,297
Upstream diversions	3,540	600	—	4,140
Delta diversions	3,781	1,099	277	5,157
West of Delta diversions	0	974	0	974
Upstream diversions	0	592	0	592
Delta diversions	0	382	0	382

SOURCE: U.S. Bureau of Reclamation 2005; Jenkins et al. 2001, Appendix F; Department of Water Resources 1998, 2005b; Dayflow data (Department of Water Resources); San Francisco Public Utilities Commission 2005; Santa Clara Valley Water District 2005; Contra Costa Water District 2005; and East Bay Municipal Utility District 2005.

NOTE: Calculations assume that consumptive use constitutes 75 percent of upstream agricultural withdrawals and 65 percent of upstream urban withdrawals. taf = thousand acre-feet.

[a]Environmental uses include net Delta outflows and water diverted to supply wetlands.

or tributary to the Sacramento–San Joaquin Delta. Because these estimates must be assembled from various sources, the particular numbers are somewhat uncertain. Nevertheless, they illustrate some important points.

First, there is little doubt that much less water flows through the Delta today than would under natural conditions.[1] In an average water year (October 1 to September 30), consumptive uses from all diversions from the Delta—about 18 million acre-feet (maf)—account for roughly 40 percent of flows that would have naturally passed through the Delta. In addition, the seasonal patterns of Delta inflows and net outflows have been altered

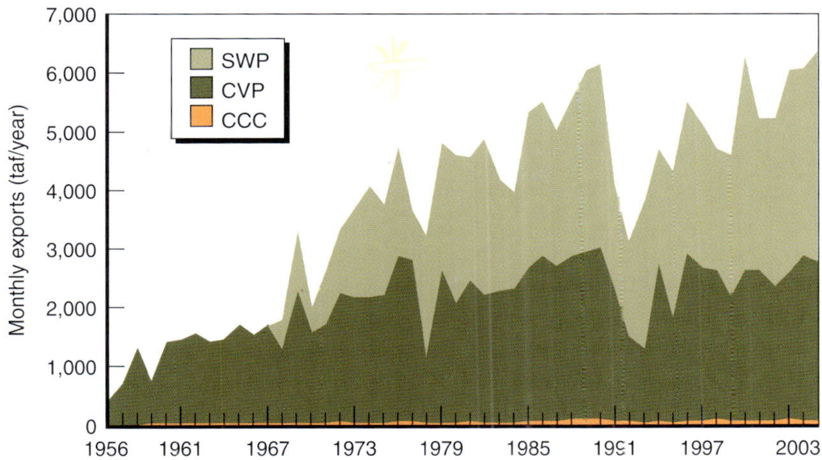

FIGURE 4.1 Major direct water exports from the Delta, 1956–2005. Totals are for water years (October 1 to September 30). Exports include the Central Valley Project at Tracy (Delta-Mendota Canal), the State Water Project at Banks (California Aqueduct and South Bay Aqueduct), and diversions for the Contra Costa Water District through the Contra Costa Canal. Source: Dayflow data (Department of Water Resources).

significantly. Today, spring Delta outflows are much less than they would be naturally, and summer outflows are generally higher.

Second, most diversions (64 percent on average) occur upstream of the Delta. To the north, Sacramento Valley water users deplete Delta inflows by almost 6.7 maf per year, mostly for agricultural uses. To the south, an additional 4.0 maf per year are consumed by diversions on the San Joaquin River and its tributaries, including the Friant-Kern Canal, which exports water to the Tulare Basin (Kern and Tulare counties). The major water projects that use the Delta as a transfer point—the Central Valley Project and the State Water Project—account for only about 31 percent of all diversions, averaging 5.4 maf per year and regularly exceeding 6.0 maf per year in recent years. The balance (4 percent) is accounted for by in-Delta users, primarily farmers.

Third, direct exports from the Delta have increased over time, with the exception of drought periods (Figure 4.1). Although exports to the federal Central Valley Project have decreased somewhat in recent years as a result of environmental flow requirements of the CVP Improvement

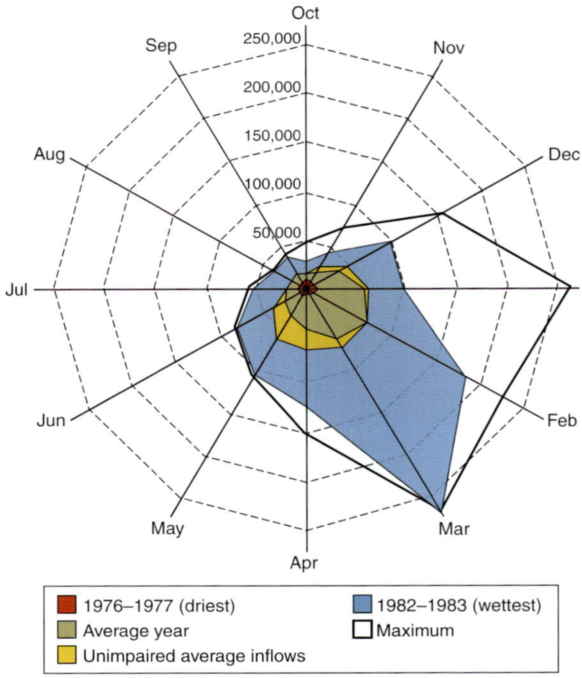

FIGURE 4.2 Seasonal and annual variability of Delta inflows (cfs), 1956–2005. Source: Dayflow data (Department of Water Resources).

Act (CVPIA), State Water Project exports have increased in response both to growth in urban water demand in Southern California and the Bay Area and to several recent wet years.

Given anticipated population growth over the coming decades, California's urban water demand is likely to increase, although conservation programs will slow the pace of this growth. However, agricultural water uses are likely to decline somewhat in reaction to market forces, including land development (Department of Water Resources 2005b). Some agricultural lands south of the Delta also will cease production as their soils become too saline to farm profitably. Some growth in urban water demands will be offset by these declines in irrigation, as well as by improvements in water conservation (particularly outdoor water conservation, which has higher consumptive water use). On balance,

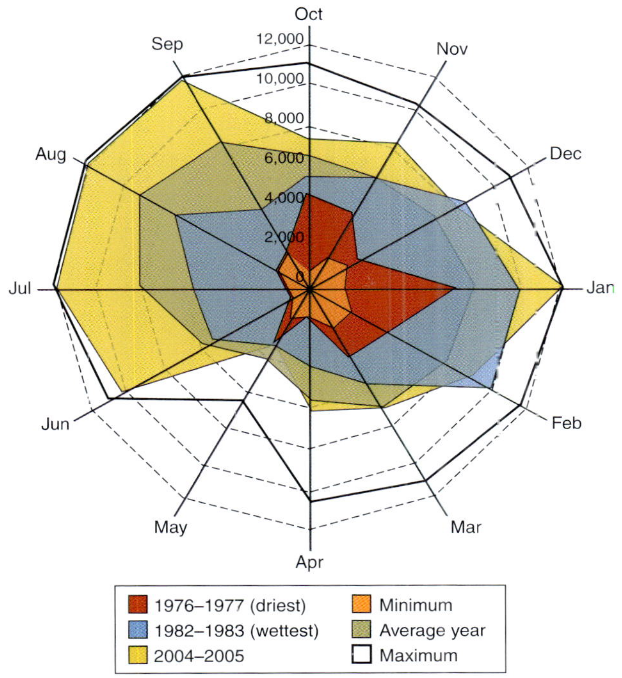

FIGURE 4.3 Seasonal and annual variability of Delta pumping (cfs), 1975–2005. Source: Dayflow data (Department of Water Resources).

large increases in total water demands for urban and agricultural uses are unlikely.

Delta water supplies remain highly variable, despite substantial management of flows through reservoir storage and releases. Inflows to the Delta from upstream sources vary greatly across seasons and years (Figure 4.2). The driest year of record (1976 to 1977) had little inflow, averaging only 2,800 cubic feet per second (cfs) for the year, and little absolute seasonal variability, ranging from 1,600 to 5,000 cfs. The wettest year of record (1982 to 1983) had an average inflow of 89,000 cfs, ranging from 23,000 to 267,000 cfs of monthly average flows. Other years of record had higher individual monthly flows, usually from floods. We estimate that, on average, the inflows that would have occurred if the Delta had been in its natural state (shown as "unimpaired" flows in Figure 4.2) tended to exceed current inflows, especially during spring.[2]

Direct water exports from the Delta are also variable (Figure 4.3), although to a lesser extent than inflows. There are two distinct seasons of pumping, winter and summer, with historically less pumping in spring and fall. This pattern results from high demand for irrigation water during the summer and the filling of off-stream storage in San Luis Reservoir in winter. It also reflects efforts to minimize pumping during the spring and fall months when fish are spawning. Annual export pumping since 1975 has ranged from 3,100 cfs (1976 to 1977) to 8,900 cfs (2004 to 2005).

FOUR WATER EXPORT STRATEGIES

The question of how to export water from the Delta has been debated and analyzed since the 1920s. Only four approaches exist for the management of Delta exports: (1) Continue pumping exports through the Delta (the current policy), (2) divert water upstream and convey it around the Delta through a peripheral canal, (3) combine the current through-Delta pumping strategy with a peripheral canal (so-called "dual conveyance" or "dual facility"), and (4) end exports altogether. All possible alternatives for managing water exports for the Delta, including those discussed elsewhere (Lund et al. 2007), are variants of these four strategies. This categorization of strategies makes it easier to compare the economic and environmental consequences of alternative Delta futures.

In this chapter, we describe the main elements of each alternative and highlight some key economic and environmental considerations. Later chapters delve further into these issues. Our analysis of these alternatives intentionally takes a very broad view, giving only limited attention to the many details of implementation (such as size, operation, and accompanying mitigations). Although the details will affect the ultimate performance of each alternative, many details can be worked out only after a strategy is chosen.

EXPORTING THROUGH THE DELTA

The Delta has four major water export locations (Figure 4.4). Listed here in order of their average export volumes per year in millions of acre-feet or thousands of acre-feet (taf), they are the State Water Project's (SWP) Banks Pumping Plant (3 maf), the federal CVP's Jones Pumping Plant

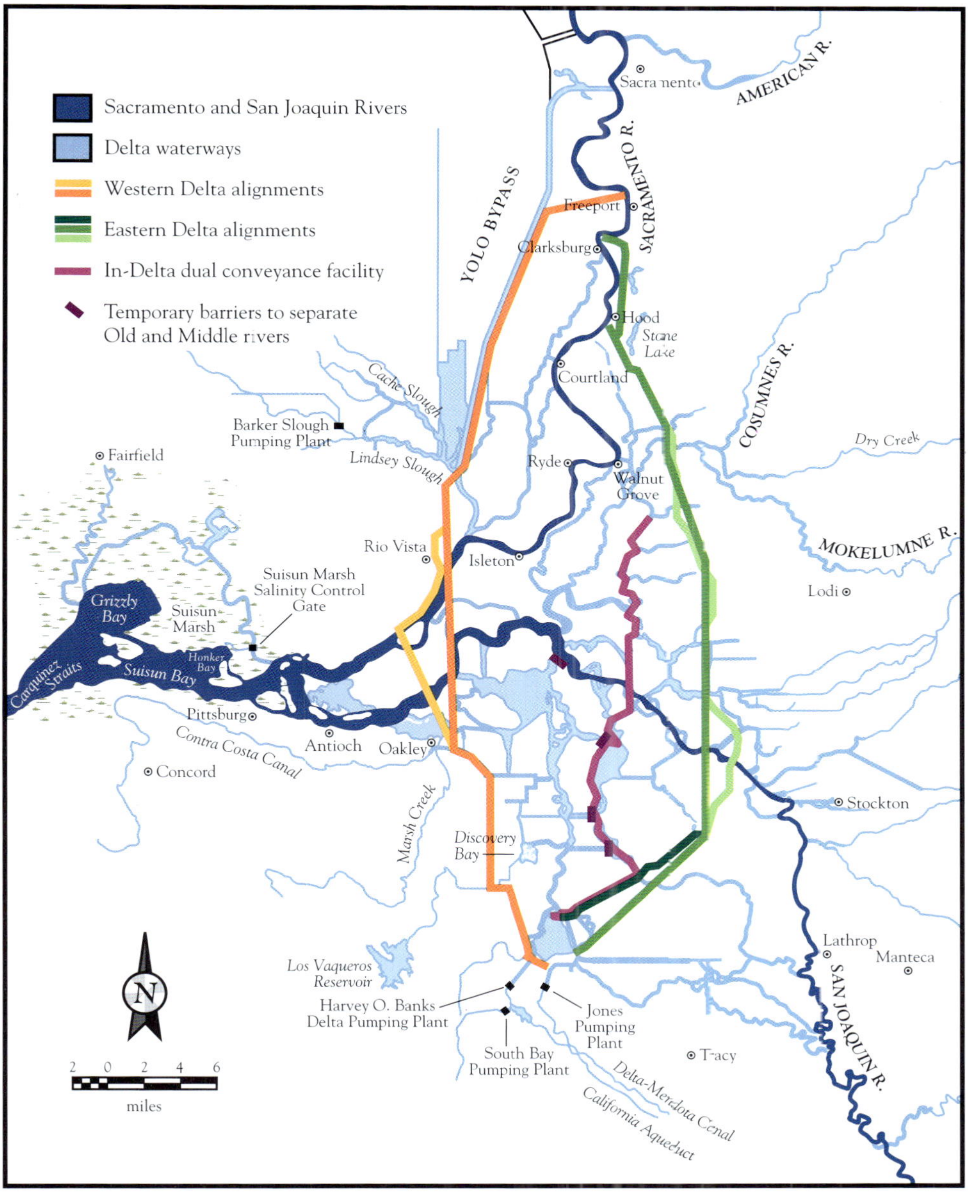

FIGURE 4.4 Some infrastructure elements for long-term export alternatives. Source: Department of Water Resources 2008.

(2.5 maf), Contra Costa Water District's federally owned Contra Costa Canal (120 taf), and the SWP's Barker Slough Pumping Plant, serving the North Bay Aqueduct (60 taf). In addition, thousands of small diversions are used to irrigate Delta farmland (totaling about 1 maf/year).

Although this water system has provided reliable water supplies for many decades, it has not been without problems. There is growing evidence that the larger points of diversion (or intakes) have harmed important aquatic species in the Delta by entraining (trapping) fish and by disrupting natural flow patterns within the Delta (see Chapter 6). At the same time, Delta water quality is becoming increasingly problematic for both human and agricultural uses (see Chapter 7).

Many alternatives have been discussed for modifying the geometry of the Delta, changing Delta inflows and pumping operations, and constructing and operating flow regulation structures to improve flows and water quality (Department of Water Resources 1957; Jackson and Paterson 1977; Orlob 1982; Lund et al. 2007). Even physical saltwater barriers have been proposed.[3] These alternatives imply a wide range of costs for construction, operations, and reductions in exports. They also would have an uncertain and potentially wide range of effects on fish. Indeed, some alternatives nearly amount to a through-Delta alignment for a peripheral canal in their degree of flow isolation from the surrounding Delta.

As sea level continues to rise and more islands permanently transition to open water, the frequency and quantity of water available from through-Delta pumping will diminish (Chapter 5).[4] This long-term loss of export capability can be reduced or delayed by diversion modifications, changes in Delta channels, and changes in operations in the Delta and upstream. Such changes will be difficult to accomplish given the potential effects on Delta fish species. Nevertheless, a series of short-term solutions might keep through-Delta exports viable for some time.

EXPORTING AROUND THE DELTA

Currently, about 11 maf per year of water is taken for "consumptive uses" upstream of the Delta in the Sacramento, San Joaquin, and eastern Delta watersheds—over 25 percent of the Delta's average natural inflows.[5] This includes such major diversions as the Tehama-Colusa Canal on the western side of the Sacramento Valley, the Friant-Kern Canal on the eastern side of the San Joaquin Valley, San Francisco's Hetch Hetchy Aqueduct, East Bay Municipal Utility District's Mokelumne Aqueduct, and many smaller surface and groundwater diversions throughout the northern and southern Central Valley. As far back as the 1920s, upstream uses were sufficiently large to affect flows and water quality in the Delta (Department of Water Resources 1995).

A peripheral canal would stop current diversions of water for export from the southern Delta, adding them instead to the upstream diversions. Many export users and Delta water users would presumably connect to the canal. Various approaches to a peripheral canal have been proposed since the 1940s (Jackson and Paterson 1977; Lund et al. 2007). Earlier peripheral canal designs were intended to expand water exports, but few today believe that a peripheral canal could be used to expand Delta exports significantly. Today's discussions include several motivations for a canal, most of which also have been considered in the past: (1) improving the quality of exported water, (2) reducing the vulnerability of water supplies to Delta levee failures, and (3) making the water supply system independent of the often conflicting demands of the Delta's ecosystem.

Moving the point of diversion of export water to an upstream location could benefit the Delta's ecosystem by ending entrainment of Delta fish and other organisms in the large southern Delta pumping plants and reducing the unnatural flow patterns in the Delta created by the pumps, which can disrupt the distribution of aquatic organisms. It would also allow more variable and environmentally beneficial management of the Delta, which would then no longer need to be kept fresh year-round to supply the pumps. However, such upstream diversions also would reduce flows of Sacramento River water into the Delta and would reduce the dilution of polluted San Joaquin River water in the southern Delta. In addition, the new intakes, if not carefully designed and managed, would entrain salmon and other fish in the lower Sacramento River. The various changes in water quality within the Delta itself, although potentially beneficial to important fish species, would pose a problem for some Delta farmers unable to connect to the canal for fresh water.

Figure 4.4 shows several types of canal alignments currently under consideration. Most analyses in the past have considered an alignment along the Delta's eastern edge (such as those shown in green). Recently, analysis of a western alignment (such as those shown in orange) has also been of interest. Although it would entail higher up-front costs (given the need to tunnel under the western edge of the Delta), such an alignment might have long-term advantages in allowing easier adjustments of the upstream intake to significant sea-level rise.

DUAL CONVEYANCE

A hybrid of the first two strategies is often referred to as dual conveyance or a dual facility (Isenberg et al. 2008a, 2008b). This approach combines elements of infrastructure and operations from continued through-Delta pumping and a peripheral canal (an eastern or western canal plus the purple through-Delta conveyance shown in Figure 4.4). The general idea is that a hybrid provides greater operational flexibility. When fish of concern are near one intake, the other intake can be used for exports. Similar benefits from redundancy might occur with respect to earthquakes, floods, and other events. With two intakes, greater overall export volumes also might be possible when large amounts of water are available, as long as near-Delta conveyance and storage capacity can accept the larger volumes. Because a dual facility relies on keeping the Delta fresh enough for the pumps, it also provides better guarantees of water quality for Delta farmers than would a peripheral canal alone.

With sea-level rise, increasing water quality concerns, and the increasing likelihood of additional permanently failed islands, the occasions when through-Delta pumping would be available are likely to diminish. This effect might be slowed by modifying intakes and Delta channels, or by increasing inflows into the Delta to improve water quality at intake locations. A wide range of proposals exist and others are being developed to allow permanent or interim use of through-Delta pumping at costs ranging from half a billion dollars to almost $10 billion.[6] Even with improvements, through-Delta pumping is likely to export less water over time. As this happens, progressively more water could be diverted through the peripheral canal. Alternatively, California could become partially weaned from dependence on Delta and Sacramento River water through a gradual, long-term reduction in exports.

ENDING DELTA EXPORTS

A final logical, if extreme, alternative is to stop exporting water altogether. Ending all exports would solve the export-related problems of fish entrainment and confusion stemming from unnatural flow patterns (see Chapter 6). Presumably, ending exports also would increase net Delta outflow from the Sacramento River watershed, something potentially beneficial for fish, although there is no guarantee that upstream users would not then take more water for use in the Sacramento Valley (see Chapter 8). Upstream users of San Joaquin River and Tulare Basin waters would face great incentives to increase diversions and transfers to supply replacement water to Southern California urban agencies and farmers in the southern Central Valley, perhaps creating additional flow and quality problems for the San Joaquin River.

Flood management, environmental flow, and water temperature management (for salmon) on the Sacramento River and its tributaries, the Trinity River, and some locations on the San Joaquin River and its tributaries would be reshaped if water exports ended. Eliminating the need to supply water for Delta exports from these reservoirs would add flexibility to release schedules for in-basin flood control, water supply, water temperature, and water quality management.

However, ending Delta exports also would impose great financial and economic burdens on water users south and west of the Delta. The costs of a catastrophic outage of water exports would be immense, and even the much lower cost of a planned weaning of water users dependent on Delta exports would be considerable (see Chapter 7). The end of water exports also would impose significant financial losses on agricultural water users north of the Delta. With the advent of water markets, many northern farmers have gained financial stability and revenues from leasing their water to users south of the Delta. Northern California's agricultural sector would lose these benefits if water exports ended.

In practice, ending exports would probably not occur at once, unless an earthquake or flood caused extensive island failures. If ending exports became a planned and phased-in policy, accompanied by expansions in conveyance connections, wastewater reuse, and perhaps seawater desalination, its costs would likely be far less than the costs of a catastrophic and unanticipated termination of exports associated with extensive and sudden failure of Delta levees.

WATER EXPORTS AND THE DELTA'S ECONOMY

The local economy of the Delta is undergoing substantial changes. Delta agriculture is likely to continue to decline and is most vulnerable and unsustainable in the most subsided areas (see Chapter 7). However, elsewhere in the Delta, agriculture is likely to remain strong indefinitely. Recreation is already a major industry in the Delta and is likely to see both change and growth with time. Total fish populations seem likely to expand as islands flood and return to aquatic habitat (Moyle 2008; Jassby and Cloern 2000). Urban development also is likely to become more important in areas where urbanization is allowed. Most of these trends are unlikely to be affected by changes in export strategy relative to other changes in the Delta brought on by the regional economy and long-term levee failures. The greatest effects of Delta export policies on the local economy are likely to be in shifting some types of recreation locally, changing types of fishing and some boating channels. But shifts in recreational activities and declines in agriculture will occur regardless of which water export alternative is chosen.

5

HYDRODYNAMICS AND THE SALINITY OF DELTA WATERS

> The invasion of salinity into Suisun Bay as far as the lower end of the Sacramento–San Joaquin Delta is a natural phenomenon which, in varying degree, has occurred each year as far back as historical records reveal.
>
> RAYMOND MATTHEW (1931), *Variation and Control of Salinity in Sacramento–San Joaquin Delta and Upper San Francisco Bay*

Since water exports began in the 1940s, the Delta has been managed to keep its water fresh enough for agricultural and urban uses by export users and in-Delta users. This management—achieved through the release of water from upstream reservoirs and changes in export schedules—can vary daily because of the Delta's complex and dynamic physical environment. Located on the eastern edge of the San Francisco Estuary and at the mouth of two major rivers, the Delta experiences numerous influences on its water quality: inflows of fresh water, saltwater, and drainage water, with substantial mixing from the tides. The future of water quality in the Delta will remain dominated by these physical forces, acting under new conditions with sea-level rise, permanent island failures, and changes in water and land management.

This chapter presents an initial examination of the implications of these new conditions for the salinity of Delta waters. Salinity is of central importance to both agricultural and urban water users. Higher salinities limit crop productivity and the ability to dispose of animal waste and raise drinking water treatment costs (see Chapter 7). More variable salinity and flow conditions than those typical under current Delta management would benefit desirable fish species (see Chapter 6). Unlike other water quality factors, such as pollutants from agricultural and urban runoff and wastewater treatment plants, salinity is likely to be strongly influenced by the physical forces acting on the Delta.

We use hydrodynamic modeling tools to examine the effects of changes in the Delta's physical conditions (sea-level rise and island flooding) and changes in water management on Delta salinity. The complexities of the Delta (multiple inflows and outflows of varying salinities, a complex network, and strong tidal influence) require numerical models to explore the effects of such changes. Although current computer models have difficulty representing sufficient physics to encompass all major driving forces of the system properly, these initial results provide important insights about future water quality in the Delta.

We find that Delta salinity is likely to increase significantly as a result of both sea-level rise and island flooding, although not all islands appear critical for keeping seawater at bay. Ending through-Delta pumping would substantially increase salinity in the southern Delta, where water quality is heavily influenced by saline drainage water from the San Joaquin Valley. Ending *all* Delta diversions, including the large volumes now diverted upstream, would make the Delta fresher overall, with greater seasonal variability. At current sea level, a peripheral canal or dual conveyance could increase salinity in some Delta locations and decrease salinity in others. Over time, however, the effects of a canal would be swamped by the increased salinity caused by sea-level rise. These findings suggest that physical forces acting on the Delta will increasingly limit the use of Delta waters for farming and urban uses, irrespective of the water management alternatives chosen. Although increased releases from upstream reservoirs and reduced export levels can help to keep the Delta fresh for some time, this continuation of the present strategy will become increasingly costly.

This investigation was performed using available hydrodynamic and water quality knowledge, supplemented with information developed with existing modeling tools. In addition to generating useful preliminary results on the effects of physical changes in the Delta and different water management alternatives, this exercise has shed light on areas where further work is needed to produce more definitive results, which we summarize at the end of this chapter. Fleenor et al. (2008) provides greater detail about this modeling work.

MODELING TOOLS AND APPROACH

We employed two existing computer models to conduct this analysis. The first is a computationally efficient, one-dimensional, tidally averaged model called WAM (water analysis module), which was developed by Resource Management Associates, Inc., for modeling work in the Delta Risk Management Strategy (DRMS)(URS Corporation and Jack R. Benjamin and Associates 2007a).[1] We use this model to explore the effects of sea-level rise and of Delta water management alternatives including the four export options discussed in Chapter 4 (through-Delta pumping, no exports, and several sizes of peripheral canal or dual conveyance facilities) and a scenario with "unimpaired flows" (eliminating both exports and upstream diversions). These simulations use historical water conditions for 20 water years, from 1981 to 2000, a period including considerable diversity in precipitation, with both one of the wettest periods and one of the longest droughts in recent history.

To explore the effects of permanent island flooding, we use results from a second, more complete model developed by Resource Management Associates, Inc., for the DRMS flooded-island modeling work (URS Corporation and Jack R. Benjamin and Associates 2007b). Because this model takes considerably longer to run (480 hours, versus 15 minutes for WAM), we relied on existing simulations for the island-flooding scenarios These simulations use conditions for a shorter period: the two and a half years between April 12, 2002, and December 31, 2004.

Unless noted, both sets of model simulations assume the same daily hydrologic and operating conditions (reservoir releases, upstream and in-Delta diversions, and exports) that occurred during the simulation periods (1981 to 2000 or April 2002 to December 2004). Both models performed well in comparisons with recorded salinity data at seven locations (Fleenor et al. 2008; URS Corporation and Jack R. Benjamin and Associates 2007a; Department of Water Resources 2007c). In addition, the flooded-island model has been successfully applied to other Delta island failures, such as the Jones Tract failure in 2004.

COMPARING SCENARIOS

In the following results, we compare Delta salinity under various scenarios with a base case representing actual conditions from April 2002 to December 2004 (for island flooding) and from 1981 to 2000 (for all other scenarios). The base case represents through-Delta pumping, with all islands intact, at current sea level. Most comparisons are shown in percentage of days each month when water salinity, as measured by the electrical conductivity (EC) of the water, exceeds a prescribed regulatory limit. To capture a range of effects for agricultural, urban, and environmental uses of Delta water, we focus on five locations within the Delta (Figure 5.1): (1) Chipps Island on the Delta's western edge—a location used to monitor salinity regulations for fish during the springtime (February through June); (2) Emmaton, a northwestern location on the Sacramento River where irrigation water standards are in effect from April through August; (3) Jersey Point, a western Delta site on the San Joaquin River (irrigation standards in effect from April through August); (4) the Contra Costa Water District's (CCWD) pumping plant in the southwestern Delta (more stringent urban standards, year-round); and (5) the Clifton Court Forebay (Clifton CF) in the southern Delta, representing exports for the State Water Project (SWP) and the Central Valley Project (CVP)(year-round urban standards and seasonal irrigation standards).[2]

Although the comparisons indicate shifts in the ability to use water for designated beneficial uses, they do not directly demonstrate regulatory compliance (or lack thereof), since the EC limits used here are fixed, whereas for most regulatory standards the limits vary seasonally and by water year type. Also, some of the regulations were not in effect during the entire period of the simulations (in particular, the environmental regulations at Chipps Island did not come into effect until 1999—see Chapter 8).

We begin by comparing through-Delta pumping with the alternatives of ending exports and ending all diversions from the Delta watershed (unimpaired flows), all at current sea level. We then explore the effects on through-Delta pumping of two physical changes in the Delta: sea-level rise and island flooding. A subsequent section examines the effects of introducing a dual facility or peripheral canal, at current sea level and with sea-level rise.

NO EXPORTS AND UNIMPAIRED FLOWS

One alternative considered here is to eliminate Delta exports altogether. Extending this logic, it is also of interest to assess the implications on Delta salinity of ending upstream diversions given that nearly two-thirds of diversions occur upstream (see Chapter 4). The role of upstream diversions in Delta water quality has been of policy interest since the early twentieth century, when the City of Antioch (at the western edge of the Delta)

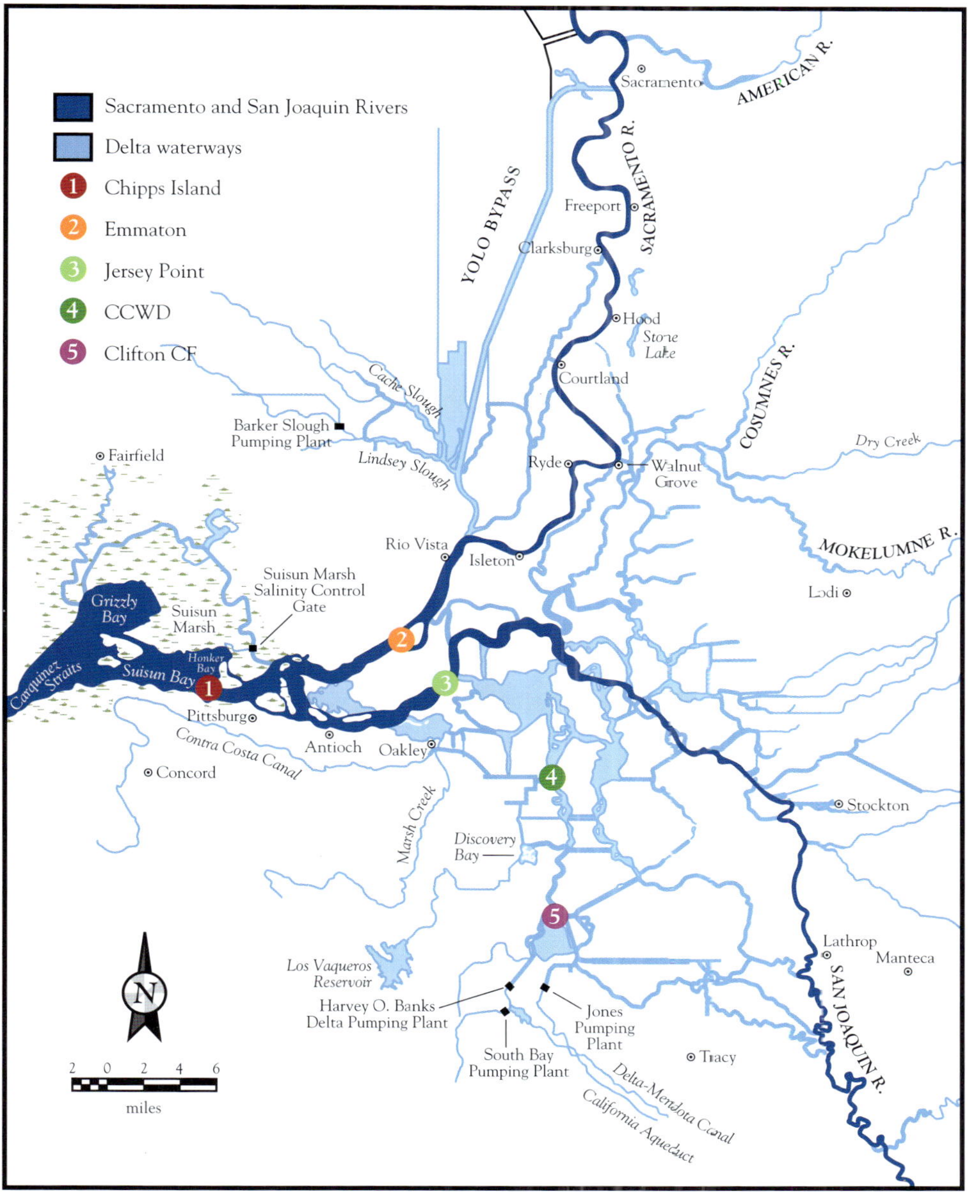

FIGURE 5.1 Locations for water salinity comparison.

unsuccessfully sued upstream irrigators on the Sacramento River for increasing Delta salinity (see Chapter 2). Most recently, the Delta Vision Blue Ribbon Task Force has emphasized the importance of considering upstream diverters when seeking a solution to the Delta's environmental woes, as described further in Chapter 8. The results of these simulations are summarized in Figure 5.2, which compares salinity conditions in the through-Delta base case to scenarios with no exports and with unimpaired Delta flows (i.e., flows without exports, upstream diversions, or surface storage).[3]

Although ending exports would substantially increase Delta outflows, it would not uniformly freshen Delta waters. Without exports salinity would decrease in the northern and western Delta (Emmaton and Jersey Point) but would increase significantly in the southern Delta (Clifton CF). (To see this in Figure 5.2, compare the position of the dark blue line, representing the base case, with the red line (no exports) and the green line (unimpaired flows) at these locations. At Jersey Point the no-exports line is absent from the figure because salinity limits are not exceeded.) Through-Delta exports reduce southern Delta salinities by diluting higher-salinity San Joaquin River water with fresher Sacramento River water drawn south to the pumps. In contrast, for unimpaired flows without upstream or export diversions or surface storage releases, dramatic reductions occur in salinity at all Delta locations, except at Emmaton and Chipps Island in the fall.[4] (Indeed, the salinity reductions are so significant at CCWD and Clifton CF that the green line representing unimpaired flows does not appear in the figure.)

Before Europeans settled the area, the Delta (and particularly the westernmost parts of the Delta and Suisun Marsh) had considerably more seasonal and cross-year variability in salinity than does today's Delta. The interior of the Delta today is now kept fresh enough to meet regulatory standards for water exports year-round, keeping Suisun Marsh and Bay more regularly saline (see Chapter 2). Greater variability can benefit the Delta's native aquatic species (see Chapter 6). The unimpaired flows scenario provides some insights into the way the Delta operated before water diversions began, although the Delta it represents is *today's* artificially modified landscape and channel networks rather than the Delta marshlands as they were before European settlement. Even in this modern Delta, salinity levels with unimpaired flows would be more variable than under export conditions (the base case). With unimpaired flows, the western Delta (Chipps Island) is fresher in the spring and more saline in the fall.[5] The intertidal, tule wetlands of the Delta before its marshlands were diked and drained would have had much more variable flow and salinity patterns across and within years. Flow rates would have been higher when water levels rose above the tule vegetation, and outflows would have been restricted at lower water levels, given the much lower natural channel capacity that existed under low-flow conditions (Baptist et al. 2007). Even with catastrophic island failures, the modern Delta would not revert to the natural Delta of pre-European times, because the islands are now highly subsided and cross channels and deeply dredged shipping channels significantly affect Delta hydrodynamics.

It is not possible to return to the Delta of pre-European times, when conditions were more beneficial for the Delta's native fish species. However, as discussed in Chapter 6 and Moyle and Bennett (2008), re-creation of more variable water flow and salinity would benefit the Delta's desirable fish. This goal is inconsistent with year-round, through-Delta pumping.

CONSEQUENCES OF SEA-LEVEL RISE

Sea level at California's Golden Gate has been increasing by 0.08 inches per year over the past century. Most climate models project an increase in the rate of sea-level rise during the next century (Intergovernmental Panel on Climate Change 2007). For planning purposes, the CALFED Independent Science Board has recommended that the Delta Vision initiative use the midrange values for sea-level rise of 8 to 16 inches by 2050 and 28 to 39 inches by 2100 (Mount 2007). Only recently have examinations begun of the consequences of sea-level rise for Delta salinity distributions.

With sea-level rise, and all other conditions unchanged, the ocean pushes its higher-salinity (higher-density) water farther into the Delta. Less clear are the potential effects of deeper water, which may reduce the vertical mixing of salinity (with fresh water at the top and more saline water at lower depths). To simulate sea-level rise, the initial water elevation throughout the Delta was increased by 1 and 3 feet. These simulations all assume continued through-Delta exports at levels experienced from 1981 to 2000.

The results, shown in Figure 5.3, compare the increase in salinity at all five locations with the "base case." With 1 foot of sea-level rise (the red line), salinity in the Delta may still be low enough for irrigation during

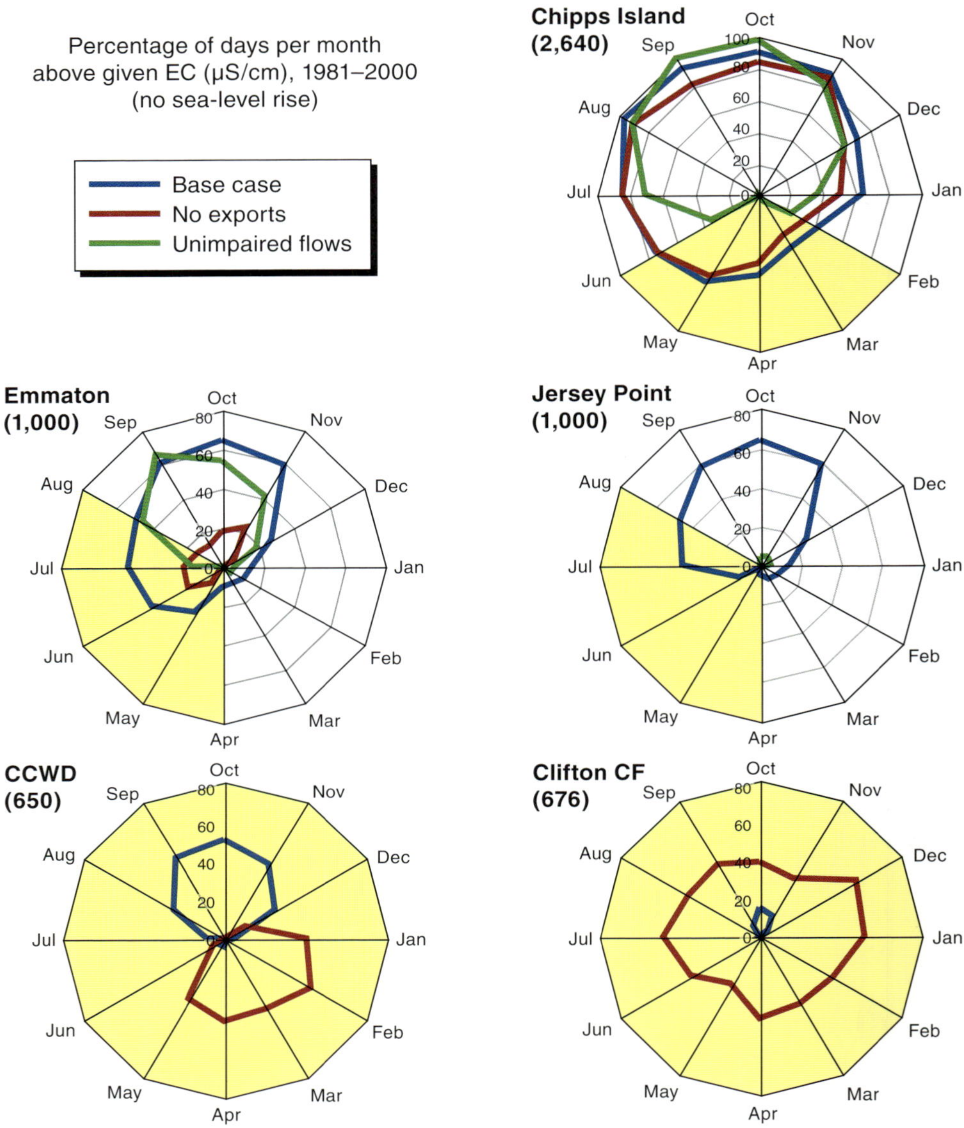

FIGURE 5.2 Effects of ending exports and upstream diversions on Delta salinity. "No exports" shows results with exports from the following locations set to zero: Contra Costa Water District (CCWD in Figure 5.1), SWP and CVP pumping plants in the Southern Delta (Banks, Jones, and South Bay pumping plants in Figure 5.1), and the SWP's North Bay Aqueduct (Barker Slough in Figure 5.1). "Unimpaired flows" also sets upstream diversions and surface storage to zero. Shaded areas are periods when compliance with salinity standards is prescribed, although compliance levels vary across water year types. In the no-exports scenario, the specified EC is never exceeded at Jersey Point. In the unimpaired flows scenario, the specified EC is never exceeded at CCWD and Clifton CF. Source: Fleenor et al. 2008.

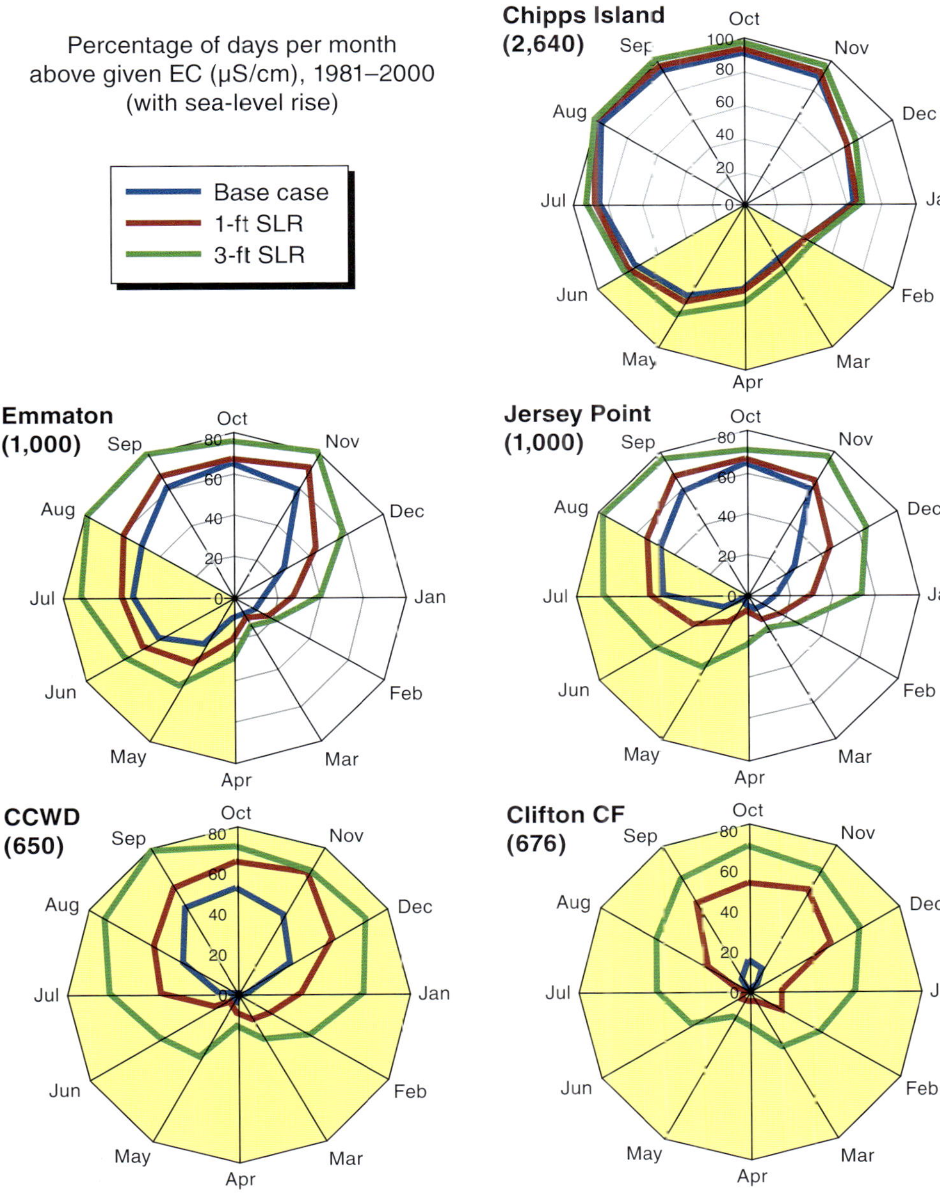

FIGURE 5.3 Effects of sea-level rise on Delta salinity. The figure shows average monthly values over the simulation period 1981–2000, with 1981–2000 levels of upstream reservoir operations and Delta exports. Shaded areas are periods when compliance with salinity standards is prescribed, although compliance levels vary across water year types. Source: Fleenor et al. 2008.

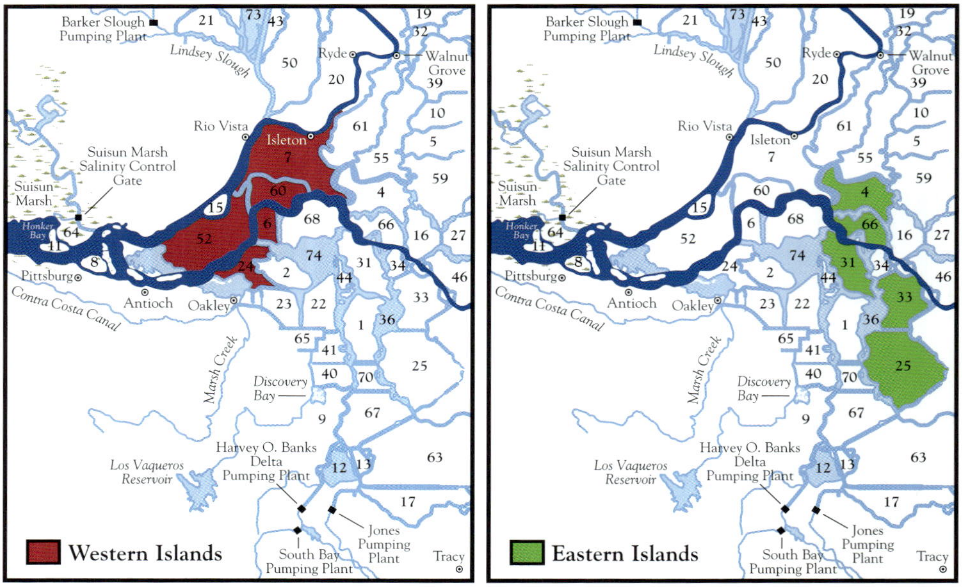

Western Islands
6. Bradford
7. Brannan-Andrus
24. Jersey
52. Sherman
60. Twitchell

Eastern Islands
4. Bouldin
25. Jones
31. Mandeville
33. McDonald
66. Venice

FIGURE 5.4 Delta island flooding scenarios. For a guide to island names, see the map at the beginning of the book (page xx).

the growing season (April to August), but higher levels in the southern Delta (CCWD and Clifton CF), particularly in the fall, would significantly increase the costs of drinking water treatment. With a 3-foot sea-level rise (the green line), salinity would greatly increase the cost of drinking water treatment and Delta water may be unsuitable for agricultural irrigation (Chen et al. 2008; Medellin-Azuara et al. 2008b). In very dry years, the salinity problems are particularly acute, even with 1 foot of sea-level rise (Fleenor et al. 2008).

One way to counteract increased salinity is to increase net Delta outflows of fresh water, by reducing export and upstream diversions and changing reservoir releases. With 1 foot of sea-level rise, an annual average of at least 475,000 acre-feet of additional Delta outflow would have

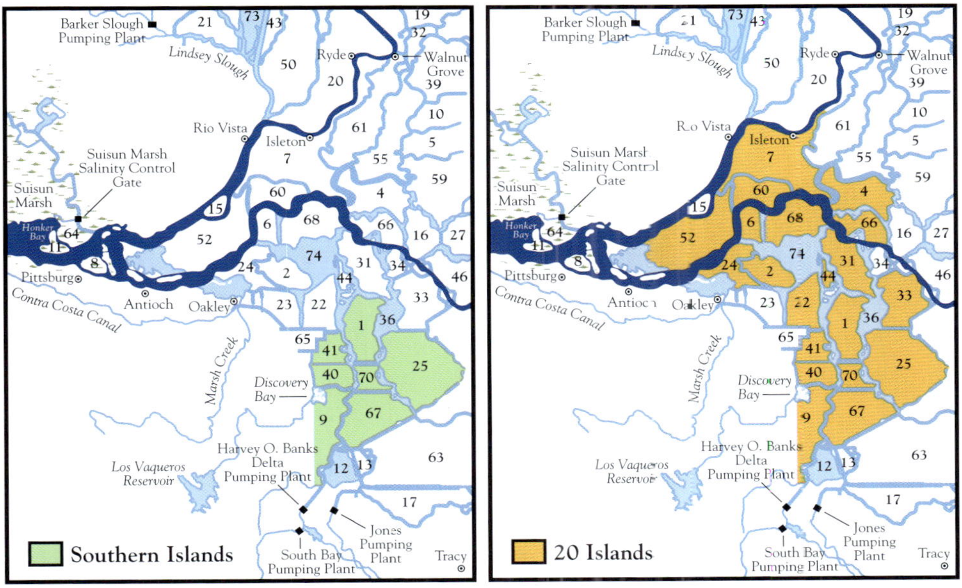

Southern Islands

1. Bacon
25. Jones
40, 41. Palm-Orwood
70. Woodward
67. Victoria

20 Islands
(all preceding islands, plus five in the Central Delta)

2. Bathel
9. Byron
22. Holland
44. Quimby
68. Webb

FIGURE 5.4 (CONTINUED)

been required to maintain 1981 to 2000 salinity conditions at the western edge of the Delta—or roughly 10 percent of annual export volumes during that period. With continued sea-level rise, the volume of required outflows would continue to increase.

CONSEQUENCES OF ISLAND FLOODING

Over the last 100 years, there have been 166 island failures in the Delta. As a consequence of continued sea-level rise, periodic flood flows, deteriorating levees, and earthquakes, islands will continue to fail; and with earthquakes and floods, many could fail simultaneously. As shown in Chapter 3 and Suddeth et al. (2008), some flooded islands may not be worth reclaiming if the judgment is based on the economic value of the activities on the islands themselves. However, it is also important to model the salinity consequences of leaving islands permanently flooded following failure, to see whether they have strategic value for keeping Delta salinity sufficiently low to continue agricultural and urban water uses.

In these simulations, the islands are "preflooded"—filled with water of salinity equaling that in surrounding channels. This depiction represents conditions for an island that has already been flooded for some time; it could also result if the initial flooding occurred during the winter or spring, when significant river flows are available. Four island-flooding scenarios (Figure 5.4) are compared with the base case, which represents through-Delta pumping as it actually occurred from April 12, 2002, to December 31, 2004, with all islands intact.

As with the other scenarios, these simulations assume continued through-Delta exports. The results, shown in Figure 5.5, suggest some striking differences in the strategic value of Delta islands for maintaining low salinity levels. The permanent flooding of five western islands increases salinity intrusion to the pumps in the southwest (CCWD) and southern Delta (Clifton CF) and would significantly affect drinking water treatment costs between August and December. In effect, the long-term consequences of permanent western island failures are almost as problematic for export water salinities as the immediate consequences of levee breaches.[6] With western island failures, more saline water is drawn in from the eastern Bay on flood tides and is then released by the islands during ebb tides into fresher river water, increasing dispersion of the salts.

In stark contrast, the permanent flooding of eastern or southern islands shows little, if any, long-term salinity effects on the Delta. There are even short periods when the failed islands improve southern Delta salinity (CCWD and Clifton CF) by facilitating the flow of Sacramento River and eastside streams (Calaveras, Mokelumne, and Cosumnes) through the Delta toward the southern pumping plants. Since the flooded areas are farther from the Bay than are the western islands, they do not draw as much saline water in from the Bay. There is also considerable eastward infiltration of sea salt from the filling and emptying of flooded western islands with each subsequent tide.

The catastrophic failure scenario (permanent failure of 20 islands) produces results similar to the failure of the five western islands, which highlights their

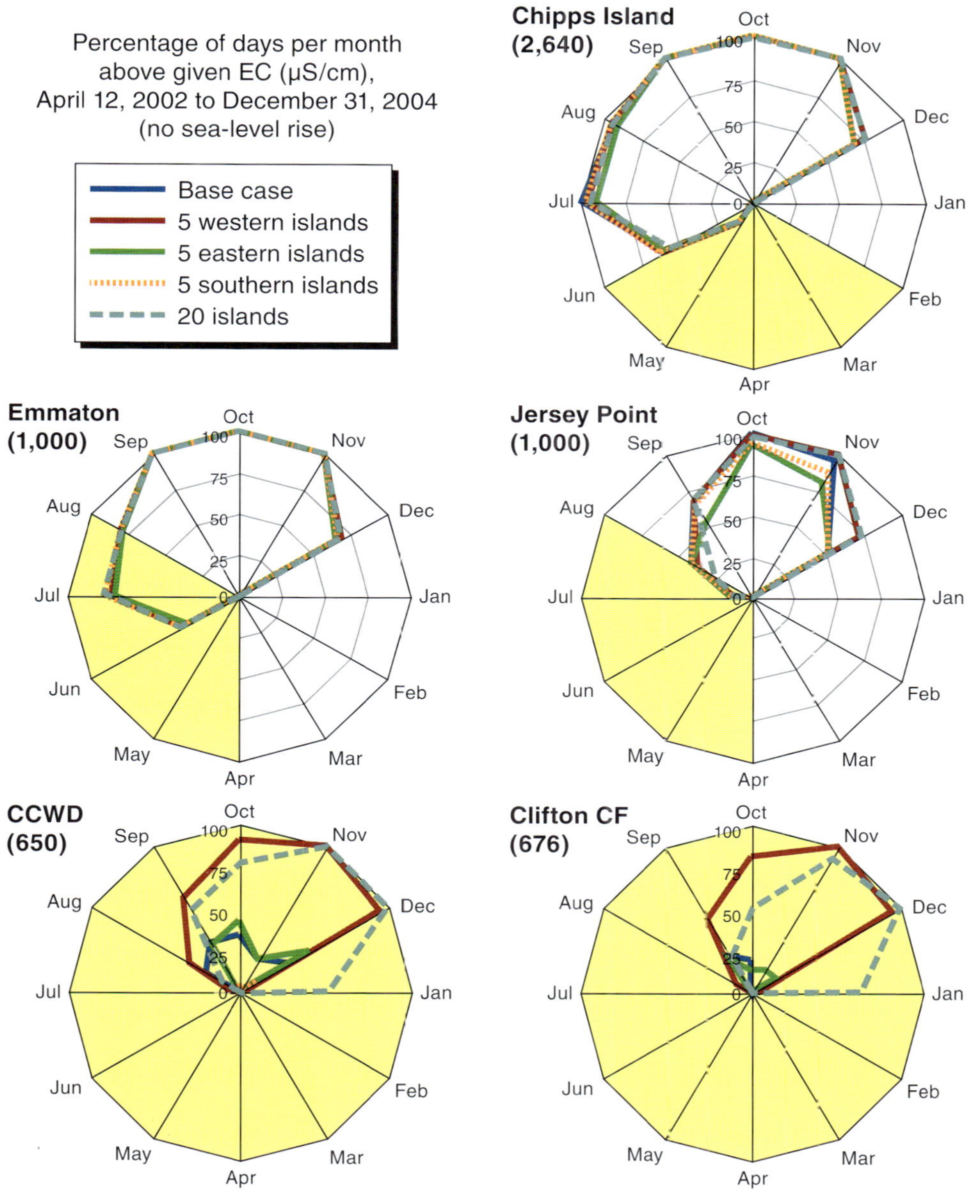

FIGURE 5.5 Effects of island flooding on Delta salinity. Shaded areas are periods when compliance with salinity standards is prescribed, although compliance levels vary across water year types. At Chipps Island and Emmaton, all five scenarios essentially overlap. Source: Fleenor et al. 2008.

importance in maintaining the current conveyance system for water exports. As discussed in Chapter 3, this is a particularly risky prospect, given the high probabilities of failure of these islands by midcentury from floods and earthquakes.

Without time available to model the combined influences of sea-level rise and island failure, we consulted others involved in modeling the Delta's hydrodynamics, who agree that the effects would at least be additive. Thus, by midcentury, Californians are likely to face conditions where large parts of the Delta have become brackish, unusable for either drinking water or agriculture without costly treatment.

CONSEQUENCES OF PERIPHERAL CANAL EXPORTS

The potential water quality effects of rerouting some or all exported water from Delta channels to a peripheral canal have been hotly debated for over 30 years. One justification for a canal has been that export users could benefit from lower salinity water by tapping into Sacramento River flows upstream of the Delta. However, water users in the Delta have been concerned that these diversions would increase salinities within the Delta itself. Although reducing or eliminating through-Delta pumping could benefit Delta fish populations, environmental advocates also have expressed concerns over whether the volume and timing of diversions would sufficiently protect fish.

The peripheral canal proposed to voters in 1982 was a large facility (a capacity of up to 25,000 cubic feet/second, cfs) intended to significantly increase water exports from the Sacramento River watershed. Here, we explore a more modest set of alternatives. We assume stability of export volumes at 1981 to 2000 levels, and we examine several canal capacities, ranging from 2,000 cfs to 15,000 cfs, operated as dual conveyance with some continued through-Delta exports. In these scenarios, the canal takes as much of the daily exports as possible, subject to an environmental constraint requiring a minimum flow on the Sacramento River of 10,000 cfs. We also examine an alternative without this environmental constraint, which operates as an exclusive peripheral canal (PC) that does not use existing Delta pumps ("PC-only"). The PC-only alternative with the same environmental constraint would require new operational patterns for exports and reservoir releases and was not examined here.

The minimum flow requirement on the Sacramento River of 10,000 cfs is introduced to prevent flow reversals, resulting from tidal influences, near potential upstream intake locations. Many organisms take advantage of tidal (bidirectional) flows, moving vertically in the water column to travel much farther on the tidal currents than they could otherwise by their own power on the downstream river current. Locating canal intakes where bidirectional flow occurs could inadvertently draw these organisms through the canal; a minimum flow requirement on the Sacramento River of 10,000 cfs has been identified as a threshold to avoid this problem (Burau 2007). Sea-level rise will move the location of the limit of bidirectional flow farther upstream. Taking water on ebb flows and not on flood tides could lessen this problem.

EXPORT QUANTITIES AND CANAL SIZE ISSUES

Table 5.1 compares the volumes of exports drawn through the canal and through the Delta for the different export alternatives and canal sizes. Although only the PC-only alternative eliminates through-Delta exports, the two largest canal capacities greatly reduce them. However, the minimum outflow constraint on the Sacramento River significantly limits the use of the canal in these scenarios. Doubling the canal capacity from 7,500 to 15,000 cfs increases average exports through the canal by less than 1,000 acre-feet per day (Fleenor et al. 2008).

However, these results should not be interpreted as justifying a hard limit on the ideal size of a peripheral canal. The scenarios examined here artificially constrain peripheral canal exports by reproducing the timing and quantity of exports that occurred between 1981 and 2000. By reoperating the system—diverting more water during high flow periods—it would be possible to export considerably higher volumes through a peripheral canal while respecting the minimum outflow requirement described earlier, as long as pumping, canal, and storage capacity were available south of the Delta. On average, the total possible deliveries through a peripheral canal with the minimum outflow constraint were 55 percent higher than the actual volume exported (7.6 maf/year possible versus 4.9 maf/year actual)(Table 5.2). Although diversions of this magnitude are likely infeasible for environmental reasons (as sharp reductions in peak Sacramento River flows would have other consequences), these results illustrate the need to consider operational

TABLE 5.1

Average Water Exports (acre-feet/day) for Four Peripheral Canal Alternatives

Alternative	*Water Export Sources*		
	Canal Diversions	Through-Delta Diversions	Total Exports
Base	0	13,500	13,500
2,000 cfs PC	3,100	10,400	13,500
7,500 cfs PC	7,900	5,600	13,500
15,000 cfs PC	8,800	4,700	13,500
PC-only	13,500	0	13,500

SOURCE: Fleenor et al. 2008.

TABLE 5.2

Potential Peripheral Canal Exports with Minimum Sacramento River Flow Requirements, 1981–2000

Export Volume	*Sacramento River Average Flow*	*Sacramento River Available Flow*[a]	*Maximum Infrastructure Capacity*[b]	*Actual Export Volumes*	*Additional Export Capacity*[c]	*Potential Additional Exports*[d]	*Total Possible PC Exports*[e]
cfs/day	24,500	16,000	14,900	6,800	8,100	3,600	10,500
af/day	48,600	31,700	29,600	13,500	16,000	7,200	20,700
af/year	17,766,000	11,590,000	10,794,000	4,948,000	5,846,000	2,626,000	7,574,000

SOURCE: Fleenor et al. 2008.

NOTE: Peripheral canal withdrawals are limited by 10,000 cfs minimum flow requirement in Sacramento River for all cases except PC-only. Canal and through-Delta diversions may not sum exactly to total exports because of rounding.

[a]Amount available after deducting minimum flow requirement (10,000 cfs).

[b]Maximum possible exports to points south and west of the Delta through the Banks (10,300 cfs) and Jones (4,600 cfs) pumps.

[c]Additional channel capacity ("Maximum Infrastructure Capacity" minus "Actual Export Volumes").

[d]Minimum of "Additional Export Capacity" and "Sacramento River Available Flow" (calculated daily).

[e]"Actual Export Volumes" plus "Potential Additional Exports."

changes and their effects on the environment before determining export capacity.

There are also environmental reasons for building a larger-capacity peripheral canal to export the same amount of water. Properly managed, a larger facility would enable water to be exported on ebb flows, during higher river flows, and only at times with less risk of environmental harm. In Chapter 8, we address some possible management approaches that could provide the necessary safeguards. Lund (2008) discusses some design safeguards for fish in the Sacramento River, using physical designs, screens, multiple inlets, and operating policies.

WATER QUALITY IMPLICATIONS OF A PERIPHERAL CANAL

At current sea-level, the dual conveyance scenarios examined here have relatively modest effects on salinities in the Delta (Figure 5.6). Salinity increases for locations along the Sacramento River (Emmaton), as the reduced river flow allows brackish water to move upstream. Salinity decreases slightly for locations near the San Joaquin River outlet (Jersey Point), as less saltwater is pulled from the west with reduced through-Delta pumping. Only the PC-only scenario significantly increases salinity at the southwestern (CCWD) and southern (Clifton CF) pumping locations, for reasons similar to the no-exports scenario examined above; with less fresh Sacramento River water being drawn toward the pumps, southern Delta water salinity is dominated by the more saline San Joaquin River flows.

For users of export water, the water salinity implications of these changes depend on the export blend, because Sacramento River water is so much fresher than San Joaquin River water. At current sea level, continued through-Delta exports with the dual conveyance systems depicted here protect agricultural users in the western Delta along the San Joaquin River (Jersey Point) and the southern Delta (Clifton CF) as well as urban users at the CCWD pumps. However, some additional upstream flow releases would likely be required to maintain agricultural salinity standards at Delta locations along the Sacramento River (Emmaton).

WATER QUALITY WITH SEA-LEVEL RISE AND A PERIPHERAL CANAL

Conditions are likely to change significantly with sea-level rise. Figure 5.7 shows the effects for export users of mixing different proportions

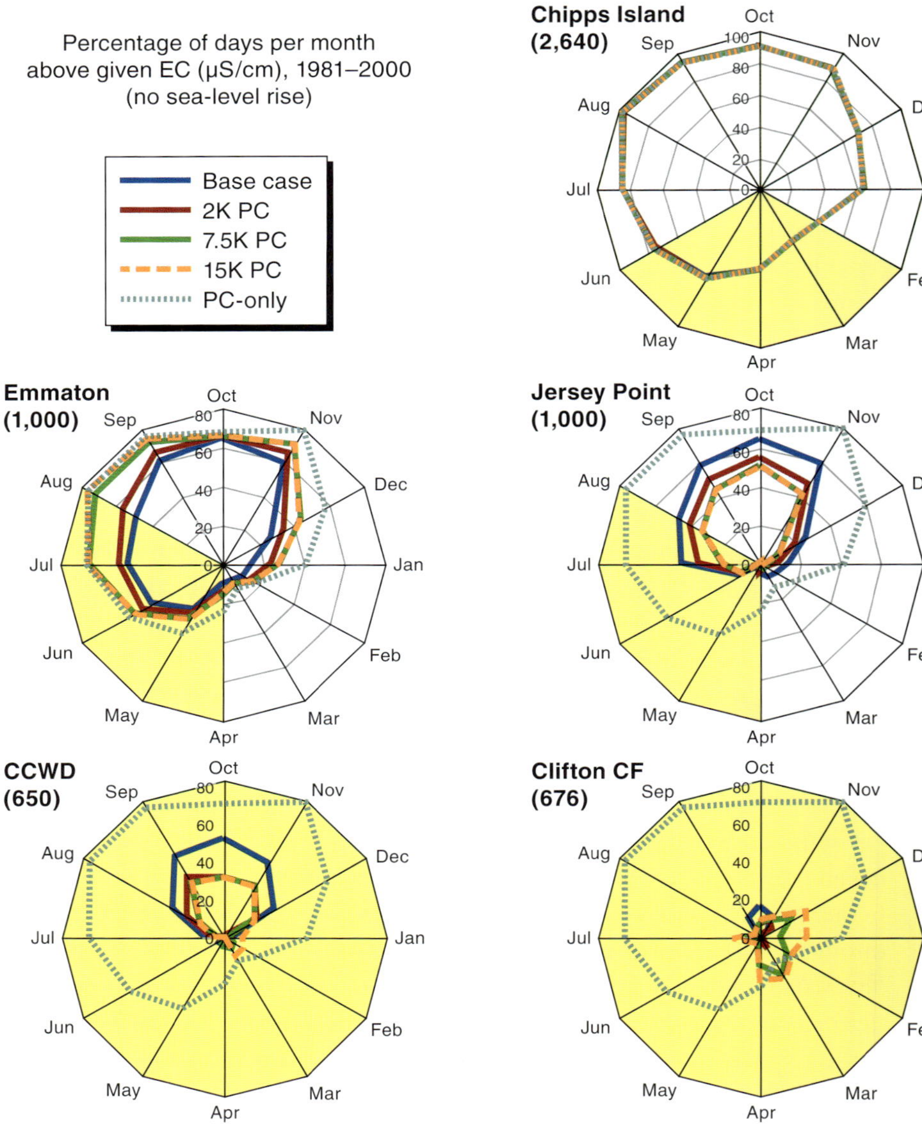

FIGURE 5.6 Effects of a peripheral canal on Delta salinity at current sea level. Dark blue line shows results of current Base Case pumping; other colors show results with the following amounts of peripheral canal capacity and a 10,000 cfs minimum flow on the Sacramento River: 2,000 cfs (red), 7,500 cfs (dark green), 15,000 cfs (orange). Light blue, hatched line show results of the PC-only scenario (15,000 cfs with no limit on diversions from Sacramento River). All scenarios overlap at Chipps Island since net Delta outflow does not change. Unshaded areas are periods where compliance is prescribed, although compliance level vary across water year types.
Source: Fleenor et al. 2008.

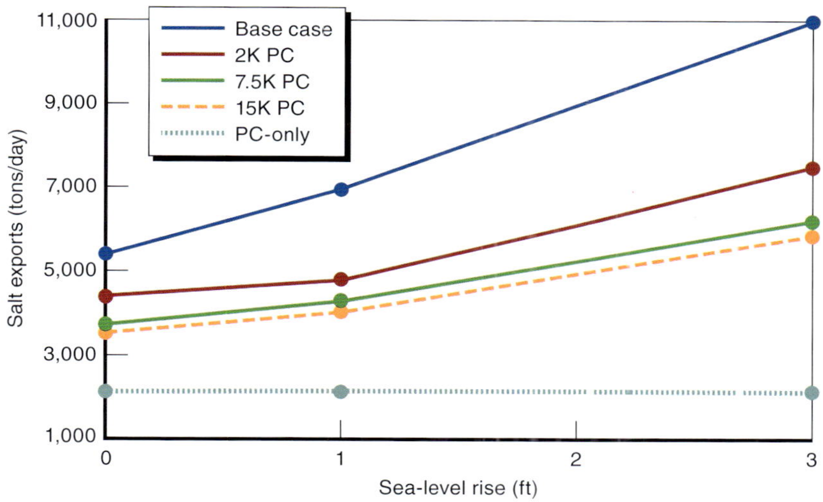

FIGURE 5.7 Salt exports for different peripheral canal capacity and sea-level rise scenarios. The calculations assume 13,547 af/day of exports on average (the 1981–2000 average). Source: Fleenor et al. 2008.

of Sacramento River and Delta water with up to three feet of sea-level rise. The figure summarizes these effects in terms of the average daily volumes of salt exported in this mix. The results confirm that export water quality deteriorates significantly with sea-level rise under the current through-Delta pumping system. A peripheral canal could significantly mitigate these effects by making fresher water available, although it would not eliminate the effects of sea-level rise if it was operated as a dual conveyance facility. Reoperation of the system (changing the timing of exports and reservoir releases) might improve these results.

To assess the effects of these changes on agricultural pumping in the Delta, Figure 5.8 shows the percentage of days during the 137-day irrigation season (April 1 through August 15) that the compliance EC levels would be exceeded at Emmaton, Jersey Point, and Clifton CF.[8] Results suggest the effects of different peripheral canal alternatives on Delta salinity will diminish over time as the sea-level rises. Under current conditions, only western Delta farmers on the Sacramento River side of the Delta (Emmaton) suffer serious salinity consequences from a peripheral canal, whereas salinity conditions actually improve for western Delta

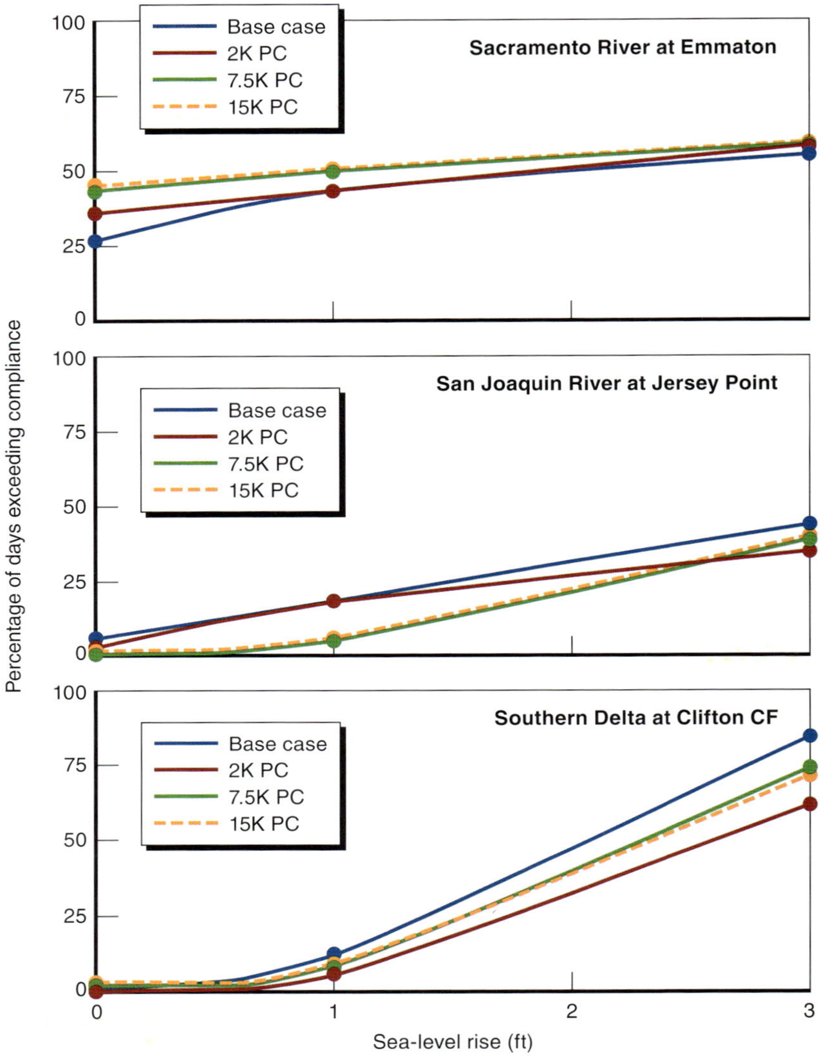

FIGURE 5.8 Average share of days above regulatory limits for irrigation, with sea-level rise. The figure shows the share of days exceeding the compliance limit for daily average EC during the irrigation season (April 1 through August 15). The results do not signify specific violations of standards because regulations are for a 14-day average. Source: Fleenor et al. 2008.

farmers on the San Joaquin River side (Jersey Point). With 1 foot of sea-level rise, conditions in the western Delta deteriorate considerably; with 3 feet, there is little difference among alternatives. Results for all alternatives suggest that with sea-level rise, irrigation in the western and southern parts of the Delta will become unsustainable in places that could not be connected to a peripheral canal.

THE LIMITS OF CURRENT KNOWLEDGE

The initial investigation undertaken here points to many areas where the knowledge base will need to be expanded to facilitate management changes in the Delta. In particular, more detailed modeling work is needed regarding sea-level rise, island flooding, and the effects of operational changes under current and future conditions. Although these gaps should not prevent California from making a strategic decision on Delta exports, it will be important to address these gaps to refine decisions about the management of flooded islands as well as environmental and water supply operations. Preparations for the Delta's future must rely on computer modeling, since little field data will available before the coming inevitable changes.

SEA-LEVEL RISE

There is some lack of agreement on whether tidal range changes with sea-level rise will be accentuated or muted in the Delta by the San Francisco Bay. In particular, if Bay Area communities erected new levees to protect infrastructure and other assets from sea-level rise (the most likely scenario), there would be a much stronger effect on the Delta than if the Bay were allowed to expand its water surface area significantly. Current models assume that the Bay geometry will remain unchanged. Expansion of the Bay's surface as a result of salt marsh restoration or abandonment of shoreline structures could lessen the effects of sea-level rise on the Delta. Better data on water depths and Bay geometry are needed to analyze these effects. This work will need to consider planning processes for San Francisco Bay, which have begun to consider coastal management decisions with sea-level rise (notably, where to armor the coastline, and where to allow inland expansion of the Bay).

ISLAND FLOODING

Although sea-level rise will eventually increase salinity in the Delta with or without island failures, additional investigation is needed to assess the minimum number of western islands required to maintain current salinity levels until the effects of sea-level rise become overwhelming. Investigations are also warranted to examine the effects of varying the locations and numbers of levee breaches. These issues challenge current modeling capabilities, which can be expanded with some steady investments.

WATER OPERATIONS

Explorations of changes in water quality regulations and operations for the new Delta will require much more in the way of hydrodynamic and operations studies than we were able to complete for this book. Model simulations also are needed to look beyond the static operations assumed in this analysis. Changes to upstream storage releases and reduced upstream diversions should be examined for different Delta export alternatives.[9] For instance, there is a need to understand the limits of canal capacity with more flexible operations of upstream reservoirs and downstream conveyance and storage capacity. Determining what volumes are feasible will depend not only on the implications of operational changes for salinity in the Delta but also on the consequences for the environment of reducing high flows on the Sacramento River below peripheral canal intakes. Operational changes also should be investigated in more detail to assess the extent of bidirectional flow changes with combined changes in operations, sea-level rise, and island failures. As an example, one solution to the increased risks of bidirectional flow effects from an upstream intake may be to take water only on ebb tides, which would require greater canal capacity.

CONCLUSIONS

This modeling exercise suggests that large changes are in store for Delta salinity as a result of natural forces acting on the Delta. Sea-level rise will eventually increase Delta salinity beyond reasonable levels for drinking water and irrigation unless large increases in Delta inflows or reductions in exports are made. Permanently flooded western islands would have a similar effect, even if the islands were preflooded to avoid a "big gulp" associated with unexpected levee failures. In contrast, islands elsewhere in the Delta might be preflooded without long-term effects on Delta salinities provided the western islands remain intact. Modeling concurrent sea-level rise and island flooding was not possible in the time available for this work. However, these two effects would at the very least be additive, making Delta water quality conditions difficult indeed for both urban and agricultural users.

Switching from the current through-Delta export pumping strategy to some form of peripheral canal or dual conveyance implies different outcomes for export water users and in-Delta water users. For export water users, a canal offers the possibility of blending in lower-salinity Sacramento River water. At current sea level, blending offers significant salinity improvements. A canal has different effects for in-Delta pumpers, depending on their location. Even when operated with minimum flow restrictions on the Sacramento River to prevent entrainment of aquatic life, a peripheral canal, operated in a dual conveyance mode, allows salinities to intrude farther up the Sacramento River, increasing salinity for Delta farmers in these areas. In contrast, salinities in the lower San Joaquin River and the western Delta generally decrease as less water is drawn into the Delta from the saltier Suisun Bay. A pure peripheral canal, without through-Delta exports, substantially increases salinity in the southern Delta, because fresher Sacramento River water is no longer drawn into the Delta to dilute San Joaquin River outflows. For southern Delta water users, this effect is similar to ending all exports. With sea-level rise, the differences among different canal alternatives diminish.

Although today's Delta bears little resemblance to the lowland marshes of the Delta in pre-European times, the modeling also confirms that salinity would be more variable if it were not managed for through-Delta pumping. Active changes in water operations would be needed to return to more variable aquatic conditions—an important facet of rebuilding the Delta ecosystem, as described in the next chapter.

6

WHAT A CHANGING DELTA MEANS FOR THE ECOSYSTEM AND ITS FISH

> In the undisturbed state of a century ago about three-fifths of the delta was awash with an ordinary tide. Spring tides could submerge all of the backswamp. River floods were capable of overflowing the entire delta, particularly when crests, high tides, and westerly winds created a congestion above the outlet into Suisun Bay.
>
> JOHN THOMPSON (1957), *The Settlement Geography of the Sacramento–San Joaquin Delta, California*

"The Delta ecosystem and a reliable water supply for California are the primary co-equal goals for a sustainable Delta." This is the first recommendation in the long-term vision for the Delta suggested by Governor Schwarzenegger's Blue Ribbon Task Force (Isenberg et al. 2008a). A major challenge to achieving such a balance is that ecosystem water demands are neither straightforward to gauge nor constant across or within years. Simply allocating some fixed proportion of the water for ecosystem purposes is unlikely to recover populations of desirable species,[1] as evidenced by the failure of "environmental" water provisioned over the last decade or more to reverse the declines of endangered fish populations. Owen (2007) points out that a key reason for this failure is that once a minimal allocation of water is made for environmental purposes, the rest is regarded as available for diversion, making it difficult to provide additional water for unanticipated needs of fish. Moreover, evaluations of the beneficial effects of special water allocations for fish have rarely been adequate. This "managing by guessing" has a well-worn track record of failure (Ludwig et al. 1993). In the Delta, this allocation strategy has resulted in a diminishing proportion of the water being made available for fish or ecosystem needs (except since 2007, when pumping was reduced to address declining delta smelt), and poor flexibility in its use, a kind of nonadaptive management (Parma and National Center for

Ecological Analysis and Synthesis Working Group on Population Management 1998).

In this chapter, we address whether alternatives to the present through-Delta pumping strategy can make the Delta ecosystem more hospitable to desirable fish and other organisms.[2] We start by discussing the basic concepts and premises that must underlie any rebuilding of the Delta ecosystem and then assess the likely responses of key species to general export strategies. We conclude with a brief discussion of initiatives needed to make it possible to manage the Delta as a resilient ecosystem that maintains desirable characteristics, as it adjusts to natural and human-induced climatic variability. Supporting analyses and background material for this chapter are found in Moyle and Bennett (2008).

BASIC PREMISES FOR REBUILDING THE DELTA ECOSYSTEM

The inevitable large-scale changes described in earlier chapters will increase tidal open-water and marsh habitat area in the Delta and decrease the area devoted to agriculture and terrestrial habitat. Although the future configurations of the Delta and Suisun Marsh,[3] and the rates at which they will change, cannot be predicted with detailed accuracy, both areas will provide very different habitats for fish and wildlife than they do today, with a likely substantial increase in aquatic habitat (Moyle 2008) . Thus, plans for ecosystem "restoration" must recognize that the new ecosystem will be very different from the present one as the result of changes as fundamental as those that transformed the Delta from marsh into farmland, towns, and roads in the past 150 years. This means that there is a unique opportunity to rebuild the ecosystem into one with attributes that society decides are desirable. By recognizing that the ecosystem will undergo major change with or without human intervention, it is possible to capitalize on the opportunity to at least partially choose the characteristics of the new ecosystem and to assess how to tailor management actions to promote these choices. The indecisive alternative—letting changes such as levee failures and invasions of harmful species occur haphazardly, and hoping for the best—is unlikely to satisfy anyone.

Ideally, ecosystem rebuilding will focus on creating habitat for species desired by society. However, many, perhaps most, species available to build the new ecosystem are the alien species from all over the world already established in the Delta. Many of these alien species are regarded as undesirable, whereas most remaining native species are regarded as desirable (as indicated by the listing of many as endangered species). One way or another, Californians will explicitly or implicitly make choices that affect the species that will dominate the new system by undertaking (or failing to undertake) actions related to physical structure, water management, fisheries, alien species, pollutants, and various other factors. Even though it is possible to promote desirable species, the rebuilt ecosystem will inevitably contain undesirable species as well, in an interacting mixture of native and non-native species, as does the present system (Lund et al 2007).

The likely changes in the Delta will potentially create habitats more favorable for desirable fish species than found in the present system, or at least unlikely to be worse. Moyle (2008) suggests that the increases in aquatic habitat caused by permanent flooding of the Delta's diked farmlands could be an improvement over present habitat for desirable native fish, such as delta smelt. The improvement should occur regardless of the water export alternative adopted, in part because conditions in the present Delta landscape are so unfavorable for many desirable species. Thus, if rebuilding the ecosystem is done deliberately, rather than by reacting to chance events, it is possible to tailor the environment to more often favor desirable species. The effectiveness of these actions also will depend on how well they discourage existing alien species that constrain ecosystem function, particularly Brazilian waterweed (*Egeria densa*) and overbite clam (*Corbula amurensis*), and how well they prevent new invasions (Lund et al. 2007).

THE ROLE OF HABITAT DIVERSITY

The key to promoting desirable aquatic species and maintaining them despite inevitable extreme episodes (e.g., prolonged drought or explosive invasions of new species) is building heterogeneous habitats, such as tidal marshes, with a diversity of flow and salinity conditions. This variability needs to take into account the physical, chemical, and biological requirements of the species at various stages in their life cycles and in different parts of the Delta.[4] Habitat diversity will not remove undesirable species, but it can disproportionately favor desirable ones, providing refuges in space and time that buffer groups of desirable species against adverse episodes. This idea is based on the concept of optimizing ecosystem resilience to avoid sudden shifts to undesirable dynamic states (Folke et al. 2004). It also draws on the principles of the Natural Flow Regime concept when managing regulated rivers (see Moyle and Bennett 2008). Under this concept, flows are manipulated to favor native species (e.g., by providing spring spawning flows for native fish).

Because most future scenarios for the Delta entail a loss of terrestrial habitat, rebuilding the Delta should include improving terrestrial habitat quality on the fringes and in less-subsided portions of the Delta and providing corridors to connect isolated patches. Such actions could actually increase habitat for the Delta's birds and other terrestrial species. The habitat likely to be lost to island flooding is largely marginal wildlife habitat associated with intensely farmed land; the biggest effects of its loss would be on wintering migratory bird species (e.g., sandhill cranes, waterfowl). Thus, rebuilding the Delta ecosystem presents an unusual opportunity to also create more wildlife-friendly agriculture and to improve existing areas for terrestrial species, especially those listed as threatened or endangered.

FISH SPECIES RESPONSES TO WATER EXPORT STRATEGIES

The effect of water export strategies on Delta fish species has been a thorny question for more than half a century. Nevertheless, it is important to assess the likely effects of these strategies as California considers long-term management options for the Delta. Here, we examine the likely outcomes for desirable Delta fish species of the four broad water export strategies described in Chapter 4: (1) continuing through-Delta pumping, (2) exporting water around the Delta with a peripheral canal, (3) using both through-Delta pumping and a peripheral canal (dual conveyance), and (4) ending water exports.

The problem is complex because exports are not the only major human influence; exports are imbedded in a system also influenced by the regulation of flow of the Sacramento and San Joaquin rivers by upstream dams and diversions, as well as by levee breaches, diversions by Delta farmers, clogging of channels by alien aquatic plants, various pollutants, and the operation of tidal gates and channels within the Delta. In the future, such problems will be exacerbated by island flooding, climate change, and other factors. Given these complexities, as well as the inevitable future surprises (such as new invasive species), we make qualitative comparisons of the likely performance of the four export alternatives based on the following:

1. Collective judgments of biologists and other Delta experts present at a February 2008 workshop
2. Groups of fish species (assemblages) that respond in similar ways to Delta conditions in time and space as shown by a multivariate analysis
3. Historical dynamics of key species and assemblages in relation to environmental variables as shown in an analysis of long-term trends
4. Major hydrologic factors that affect fish populations as the result of export strategies or changes in diversions

EXPERT JUDGMENTS

The results of a survey conducted in February 2008 at a workshop of the Interagency Ecological Program's Estuarine Ecology Team captured the current thinking of 39 Delta aquatic ecosystem experts (primarily biologists with more than two years' experience in working on Delta issues) about the future of key fish species (Bennett et al. 2008). Most thought that declining species such as delta smelt and Chinook salmon would likely continue to decline, with some probability of extinction, regardless of the export path chosen. However, ending exports entirely was considered the most beneficial for these fish species (Figure 6.1). Continuing with through-Delta exports alone was considered least beneficial; operating a dual conveyance facility or a peripheral canal alone tied in an intermediate position. This general assessment agrees with our own.

FISH GROUPINGS

To assess whether groups of desirable fish species are likely to respond similarly to changing Delta conditions, we conducted a multivariate analysis (principal components analysis) to determine if Delta fish species fall into

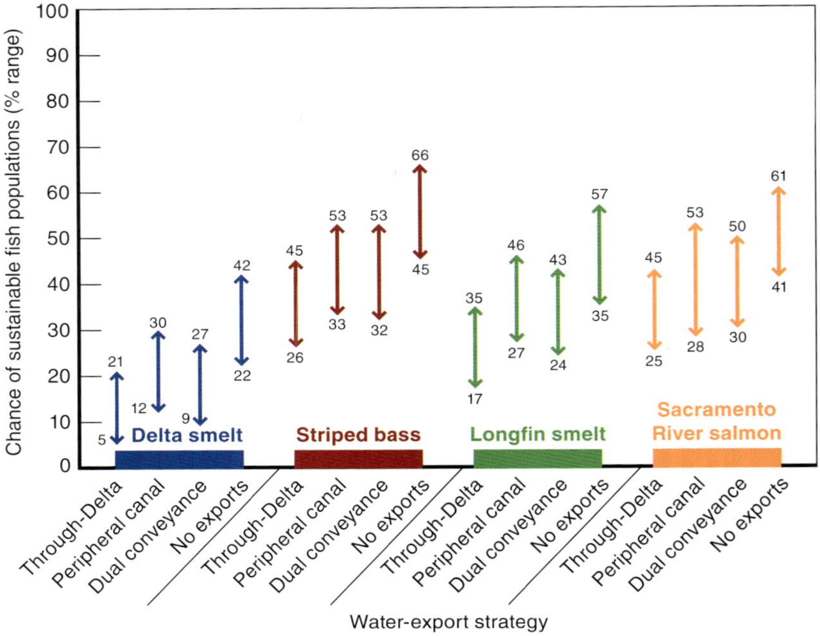

FIGURE 6.1 Expert assessment of likelihood of sustainable fish populations in the future with different water export strategies. Figure reports averages of 39 experts' high and low estimates of fish population viability several decades into the future. Source: Bennett et al. 2008.

natural groups for management purposes (Moyle and Bennett 2008). Five such groups were identified:

1. "Smelt" group (desirable)—short-lived pelagic species, including delta smelt and longfin smelt
2. "Anadromous" group (desirable)—striped bass, American shad, Chinook salmon, plus brackish-water benthic species (those that live and forage on or near the bottom), including staghorn sculpin, starry flounder
3. "Freshwater benthic" group (desirable)—mostly native species (splittail, Sacramento sucker, prickly sculpin, tule perch) plus non-native mosquitofish and shimofuri goby

4. "Delta freshwater planktivores" group—threadfin shad, inland silverside, hitch
5. "Slough-resident fish" group (mostly undesirable), which includes the centrarchids—mostly non-native species associated with beds of aquatic vegetation, including largemouth bass, bluegill sunfish, bigscale logperch, common carp, white catfish

Most fish in the last two groups either have negative interactions with desirable Delta fish by preying on them, competing for food and habitat, or do well in the areas infested with water weeds that are avoided by most of the desirable fish species. The results of this analysis suggest that the Delta can be managed to favor groups of desirable species, as discussed further later.

SPECIES DYNAMICS AND ECOSYSTEM REGIME SHIFT

To assess how key species and assemblages have responded to changing Delta conditions over the past 40 years, we conducted an analysis of trends in several key variables, including water exports, salinity, and water clarity (from July to September), and the annual biomass of key native and alien species (a measure of fish abundance)(Moyle and Bennett 2008). Figure 6.2 presents these data in phase plots, which depict the results for each variable relative to its long-term average (set to zero) for each year from 1976 to 2006. The most recent years (2000 to 2006) are depicted by solid red dots and earlier years by hollow blue dots. In general, the analysis shows that there was more variation from the long-term average in years before 2000. The reduced variation in recent years indicates a broad shift toward undesirable environmental conditions. As water exports during summer have increased, the variability in Delta salinity has decreased (panel A), maintaining lakelike conditions favored by Brazilian waterweed, which in turn creates conditions that favor undesirable fish species (mostly the slough-resident fish species, described earlier) and that are unfavorable for desirable native fish species. As the abundance of inland silverside (an alien potential predator on and competitor with delta smelt) has increased, the biomass of delta smelt has become consistently small (panel B). As salinity has become less variable during summer, water clarity has increased (as a result of declining sediment input from rivers and an increase in water weed abundance)(panel C), providing less favorable habitat for desirable

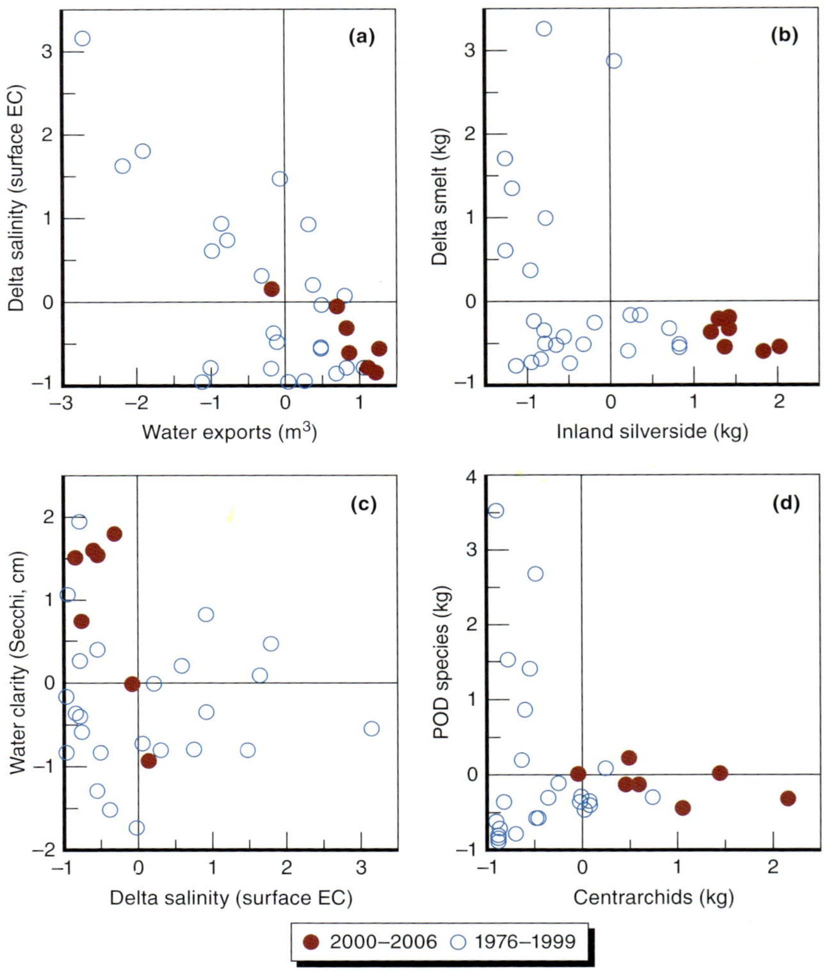

FIGURE 6.2 Regime shift in the Delta ecosystem. Data are from July through September for 1976–2007. Data are plotted with zero mean and unit standard deviation. Source: Moyle and Bennett 2008.

fish species. As alien centrarchid fish species (largemouth bass and bluegill sunfish) have become more abundant, the biomass of native smelt and striped bass (POD or pelagic organism decline species) has become smaller and less variable (panel D). The decline in variability of these conditions since 2000 suggests the Delta has entered a new ecological regime.

This analysis leads us to conclude that the Delta ecosystem has shifted toward a regime of reduced environmental variability, especially in salinity and water clarity. Essentially, the Delta ecosystem has shifted over the past decade from a regime favoring delta smelt and other native species to one favoring undesirable alien fish species.

Shifts to new regimes are thought to be triggered by the interactions between "slow" and "fast" processes once some critical threshold is exceeded (Scheffer and Carpenter 2003). The long-term rising trend in water exports (a slow process) has constrained the natural variability in flows and other environmental conditions, facilitating the proliferation of alien species, including invasion of new species (a fast process). The interrelationship of these processes appears to have tipped the system dynamics since the early 2000s. The low variability in recent years, potentially enhanced by the habitat-stabilizing properties of Brazilian waterweed and the long lifespan of the predatory largemouth bass, suggests that it will be hard to push the ecosystem back to a regime favoring desirable species without significantly altering Delta water management.

HYDROLOGY AND DELTA FISH SPECIES

The following hydrologic changes can affect Delta fish species:

- Changes in inflows, particularly from the Sacramento River
- Increased flooding of Delta islands
- Changes in the movement of water across the Delta
- Salinity intrusion in the western Delta
- Dispersion and dilution (or concentration) of pollutants (e.g., ammonium, pesticides) entering the Delta, particularly those in agricultural drainage from the San Joaquin River
- Changes in the total volume of through-Delta pumping
- Changes in the amount of water diverted upstream of the Delta, particularly with a new intake for exports on the Sacramento River

The four export management alternatives will affect these factors differently. One way to look at how the alternatives affect fish is to score each one according to its likely positive and negative effects on fish. Table 6.1

TABLE 6.1 *Estimated Effects of Export Strategies on Major Hydraulic Factors and Delta Fishes in the Next 50 Years*

Effects	*South Delta Pumping*		*Peripheral Canal*		*Dual Conveyance*		*No Exports*	
	Effect on factor	Effect on fish	Effect on factor	Effect on fish	Effect on factor	Effect on fish	Effect on factor	Effect on fish
Sacramento River inflow	0	0	−2	−	−1	−	2	+
Island flooding	−1	−	2	+	−1	−	2	+
San Joaquin drainage water	1	−	2	−	1	−	1	−
Salinity intrusion	−2	−	2	+	−1	−	2	+
Across-Delta flow	2	−	−2	+	−1	+	−2	+
Pumping from Delta	2	−	−2	+	−1	+	−2	+
Intake on Sacramento River	0	0	2	−	1	−	0	0

SOURCE: Author estimates, assisted by expert survey results (Bennett et al. 2008, Tables E.3 and E.4).

NOTE: *Effects on factor* are measured on a five-point scale: −2 indicates a major decrease in the effect of a factor, −1 indicates a moderate decrease, 0 indicates no change or a very small decrease from today's conditions, 1 indicates a moderate increase, and 2 indicates a major increase. *Effects on fish* are measured as probable directions of effects on desirable fish species: "+" (positive), "−" (negative), "0" (no change). Thus, the export strategy of south Delta pumping has a score of 2 on "pumping from the Delta," indicating that this export strategy has major effects on Delta waterflow patterns through the effects of the pumps, while a score of −2 on "salinity intrusion" indicates that this export strategy would maintain low salinity in the Delta and in exported water. Both factors would have negative effects on desirable fish.

presents such a ranking, drawing largely on our own professional judgment while taking into consideration the opinions of experts surveyed at the February 2008 workshop.

These results suggest ending Delta exports is best for estuarine fish because it would allow the resumption of more natural estuarine flow patterns, assuming no major increase in the amount of water diverted upstream from the Sacramento River and its tributaries. If hydrologic conditions remain roughly the same as today, then through-Delta pumping clearly has major effects on Delta flow patterns, with harm to desirable

TABLE 6.2

Potential Effects of Water Export Strategies on the Five Groupings of Delta Fishes

Fish Species Group	*South Delta Pumping*	*Peripheral Canal*	*Dual Conveyance*	*No Exports*
Smelt (desirable)	−2	1	−1	2
Anadromous (desirable)	−1	−2	−1	1
Freshwater benthic (desirable)	0	1	1	1
Delta freshwater planktivores	0	0	0	1
Slough-resident fishes	2	1	1	0

NOTE: Effects are measured on a five-point scale: −2 = strong negative change from present conditions (decrease in distribution and abundance), −1 = moderate negative effect, 0 = no effect or small effect, 1 = moderate positive effect, 2 = strong positive effect.

fish species, especially by entrainment. The likely effects of a peripheral canal on fish are strong but mixed. Although a canal largely eliminates the negative effects of through-Delta pumping, the large diversions on the lower Sacramento River would reduce inflows to the Delta and might have major effects on salmon and other fish that swim upriver. It is also unclear whether fish that live in the Delta would be harmed by the greater influence of low quality San Joaquin River water, with its warmer temperatures and high salt and pollutant loads, or whether higher nutrient inputs from this water would stimulate Delta food webs and eventually benefit desirable species. Ecosystem function would depend largely on the amount and timing of Sacramento River water allowed into the Delta, as well as on the extent of island flooding. Dual conveyance retains the generally negative effects of through-Delta pumping, but these disadvantages could be ameliorated somewhat by the more opportunistic pumping afforded by dual intakes.

Because some of these hydrologic influences affect different fish differently, it is useful to consider the likely effects of the export strategies on each of the five groupings of Delta fish species identified earlier (Table 6.2).

Continuing through-Delta pumping would be generally bad for the three groupings of primarily desirable fish species. Because the status quo has demonstrably negative effects on desirable fish species, it is unlikely that just tinkering with operations of this water delivery system will make things much better.

Relying entirely on the peripheral canal for water exports might be better than the present system for pelagic fish species such as delta smelt and longfin smelt. The two most compelling drawbacks are (1) the eggs and larvae of striped bass and American shad may face increased entrainment from an upstream intake, and (2) juvenile Chinook salmon may encounter entrainment problems with the intake and increased water quality problems on the San Joaquin River side of the Delta. A canal could be designed and operated to greatly reduce these threats to fish by better aligning the timing and amounts of water entering the Delta to foster desirable ecosystem attributes. In this case, the score of a peripheral canal–only alternative would improve in Tables 6.1 and 6.2.

The dual conveyance alternative has the potential for more flexible management than does a peripheral canal–only alternative (see Chapter 4). Although dual conveyance in theory could reduce the negative effects on desirable fish species by allowing switching between the two intakes, it could also be operated in ways that combine the negative effects of both the peripheral canal and through-Delta pumping. Dual conveyance would need to be operated carefully to avoid direct entrainment effects and dramatic reductions of freshwater inflows to the Delta.

The no-exports alternative would have generally positive effects on all three groups of desirable fish species. Although it may be unrealistic to implement this option because of costs to human water users, its consideration, from an ecosystem perspective, provides a model for creating better conditions for desirable fish species. However, even the positive effects of this option can be overridden by invasions of new alien species with ecosystem-altering properties, as has happened in the recent past with overbite clam and Brazilian waterweed. Like all other options, it would have to be accompanied by major efforts at habitat restoration to be really successful.

WORKING WITH UNCERTAINTIES

The results of these comparative analyses of export options, although crude, indicate the difficulties of exporting large amounts of water from the Delta in a way that is compatible with fish conservation and the rebuilding of an ecosystem with many desirable attributes. If an improved Delta environment is to be a major societal goal, then a major effort must be made to create that environment by working *with* the inevitable large-scale changes, rather than fighting them, or ignoring them (only to lose). There must be enough flexibility with diversions and inflows to reduce direct entrainment effects on fish species and allow for variable environmental demands.

The projected large-scale changes to the Delta will irreversibly alter the state of the system. The new Delta can have attributes that favor desirable fish species (and other organisms), such as maintaining Suisun Marsh as a tidal brackish system, open-water habitat relatively free of overbite clam, Brazilian waterweed, and other invasive species, with a more natural tidal and inflow pattern. However, creating an ecosystem with desirable attributes first requires societal recognition that humans benefit from the ecosystem services provided by this highly altered system.[5] Only then can resiliency to future change be built into the new system (Scheffer et al. 2001; Folke et al. 2004; Walker and Salt 2006).

Unfortunately, not all aspects of the inevitable large-scale change can be ameliorated by new management practices. The rise in temperature of the freshwater inflow as the result of climate change is particularly worrisome because it will increase the stress on native fish species, especially delta smelt. Delta smelt survival in the face of this change would be possible if the species can adapt to the changing temperature regime or if the inevitable increase in river temperatures will be countered by intrusion of cooler tidal waters and ensuing changes in estuarine hydrodynamics.

ATTRIBUTES OF AN ECOSYSTEM SOLUTION

To tackle the problems of the Delta, it is always important to keep in mind that the rebuilt ecosystem will contain constituents and characteristics far removed from those that existed previously. The historical Delta was unique in its characteristics, as the future Delta will be, with only superficial resemblance to current and past Deltas. If reliable exports and a healthy environment are to be coequal goals, it will be necessary to work systematically to foster favorable ecosystem attributes, while recognizing that human control over natural events is limited. Figure 6.3 provides an illustration of what an eco-friendly Delta might look like in the future.

Next we describe some potential actions to achieve the type of ecosystem solution depicted in Figure 6.3. These are presented as actions to guide

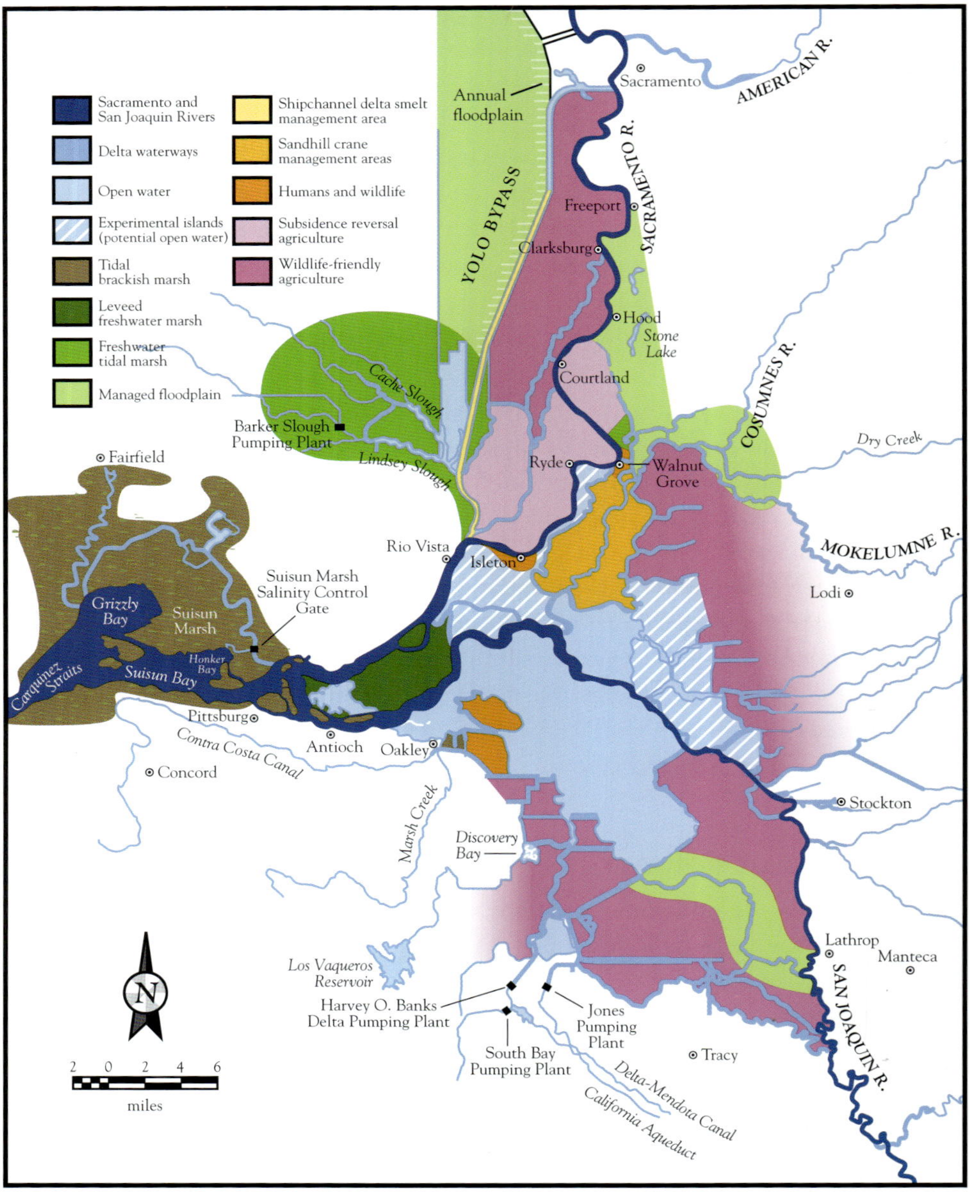
Sacramento and San Joaquin Rivers
Delta waterways
Open water
Experimental islands (potential open water)
Tidal brackish marsh
Leveed freshwater marsh
Freshwater tidal marsh
Managed floodplain
Shipchannel delta smelt management area
Sandhill crane management areas
Humans and wildlife
Subsidence reversal agriculture
Wildlife-friendly agriculture
Annual floodplain
Sacramento
AMERICAN R.
SACRAMENTO R.
YOLO BYPASS
Freeport
Clarksburg
Hood
Stone Lake
Courtland
COSUMNES R.
Dry Creek
Cache Slough
Barker Slough Pumping Plant
Lindsey Slough
Ryde
Walnut Grove
MOKELUMNE R.
Fairfield
Rio Vista
Isleton
Lodi
Suisun Marsh Salinity Control Gate
Grizzly Bay
Suisun Marsh
Honker Bay
Suisun Bay
Carquinez Straits
Pittsburg
Antioch
Oakley
Contra Costa Canal
Concord
Stockton
Marsh Creek
Discovery Bay
Lathrop
Manteca
Los Vaqueros Reservoir
Harvey O. Banks Delta Pumping Plant
Jones Pumping Plant
South Bay Pumping Plant
Tracy
SAN JOAQUIN R.
Delta-Mendota Canal
California Aqueduct
N
2 0 2 4 6
miles

FIGURE 6.3

solution-oriented thinking, as they are applied in several key areas: planning and regulation, experimentation, toxicants, flow patterns, and habitat restoration.

PLANNING AND REGULATION

- Develop a coordinated, system-wide vision and plan for ecosystem rebuilding, with management decisions guided by modeling and experiments in large-scale environmental manipulations (i.e., active adaptive management).
- Develop an aggressive regulatory system with tools to help prevent new invasions of alien species to prevent them from reconfiguring desirable ecosystem attributes (as Brazilian waterweed and overbite clam have done in the past decade).

EXPERIMENTATION

- Conduct large-scale environmental experiments to learn how to better manage the new Delta's habitats. Potential experiments include (1) flooding one island while monitoring closely the effects on aquatic organisms and water quality; (2) constructing a gated, floodable island on the Delta Wetlands model that can be used to experiment with different flooding, salinity, and temperature regimes;[6] and (3) testing large-scale programs to reduce populations of alien species that are ecosystem engineers (e.g., removal of overbite clam by dredging).

FIGURE 6.3 Land and water use in an eco-friendly future Delta. The key aspects of this map include: (1) protecting levees in the western Delta to allow for at least opportunistic through-Delta pumping; (2) large expanses of pelagic, open-water habitat; (3) large areas maintained for environmentally friendly agriculture; (4) Suisun Marsh recreated as a brackish water tidal marsh; (5) large areas of freshwater tidal marsh; (6) the Sacramento ship channel and deep areas of Cache Slough managed for delta smelt spawning; (7) large expanses of floodplain, with annual floodplain created along the eastern edge of the Yolo Bypass; (8) the Stockton ship channel maintained through a larger area of open water (shown here as the San Joaquin River); (9) the integrity of the Sacramento River maintained through the Delta for salmon migration; and (10) islands reserved for experimental use, including flooding.

TOXICANTS

- Reduce pollutants, especially ammonium and pesticides, from regional agriculture and urban areas through education, incentives, better regulation, land retirement, and cap-and-trade systems for limiting and allocating pollution loads.[7]
- Reduce or eliminate input of salts, warm water, and toxic pollutants to the southern Delta from the lower San Joaquin River.

FLOW PATTERNS

- Create a Delta inflow pattern flexible enough to adjust to changes in physical ecosystem structure (e.g., island flooding) while providing needed flows for desirable species to successfully complete their life histories (e.g., for spawning of delta smelt).

HABITAT RESTORATION

- Create diverse, native fish-friendly tidal habitats on the peripheries of the Delta, especially in the Cache Slough region and Cosumnes River. (For the Cache Slough region, this would include relocation of the North Bay Aqueduct intake.)
- Recognize that much of Suisun Marsh will become subtidal and tidal brackish water habitat as the result of sea-level rise. Attempt to address the change through management of levees and areas most likely to flood.
- Improve the Yolo Bypass as a fish habitat, especially through the annual flooding of some areas; create a similar bypass on the San Joaquin side (e.g., on Stewart Tract).
- Create large blocks of upland habitat on the margins of the Delta and create corridors to connect isolated patches of habitat, to favor terrestrial species, especially overwintering birds. Much of this land could be devoted to wildlife-friendly agriculture.

CONCLUSIONS

This assessment of the long-term prospects for the Delta ecosystem and its fish species leads to five main conclusions.

First, large-scale ecosystem change is inevitable in the Delta and the new ecosystem will be very different from both the historical and the present ecosystems. Key aspects of the new ecosystem will be large areas of open-water habitat and greater heterogeneity in environmental conditions, especially salinity.

Second, a general factor associated with the decline of desirable fish species in the Delta in the past two to three decades has been a trend for the environment to become a less heterogeneous, more freshwater-dominated system. The low variability in summer water quality, potentially enhanced by the habitat-stabilizing properties of Brazilian waterweed, suggests it will be hard to push the ecosystem back to a regime favoring desirable species without significantly altering Delta water management.

Third, different groups of fish species are favored by different sets of environmental conditions, indicating that general management strategies can be established to disproportionately benefit groups of desirable species (e.g., pelagic and anadromous species). However, the benefits of any management strategy can be greatly reduced by the invasions of new alien species.

Fourth, the best water export strategy to favor desirable fish species is to end exports, assuming that upstream diversions do not increase substantially. The worst strategy is to keep pumping large amounts of water through the Delta. Any export strategy (including ending exports) must include a large component of restoring habitat diversity and function throughout the Delta and Suisun Marsh if it is to be successful at bringing back large populations of desirable fish species.

Fifth, because of high uncertainties as to how ecosystem change will affect desirable species, large-scale *in situ* experiments are needed (e.g., flooding islands) to find management strategies that have the highest likelihood of success. In addition, several large-scale restoration projects identified here (Cache Slough, Suisun Marsh, and Yolo Bypass) are very likely to benefit many desirable species.

7

ECONOMICS OF CHANGING WATER SUPPLY AND QUALITY

> The rains of California are ample, but confined to Winter and Spring. In time, her streams will be largely retained in her mountains by dams and reservoirs, and, instead of descending in floods to overwhelm and devastate, will be gradually drawn away throughout the Summer to irrigate and refresh. For a while, water will be applied too profusely, and injury thus be done; but experience will correct this error; and then California's valleys and lower slopes will produce more food to nourish and fruit to solace the heart of man than any other Twenty Millions of acres on earth.
>
> HORACE GREELEY (1868), *Recollections of a Busy Life*

The Delta is a major source of water for urban and agricultural uses in the Bay Area, the southern Central Valley, Southern California, and the Delta itself. The recent rise of water markets has more closely linked water management in upstream and importing regions of the state, and the evolving natural conditions in the Delta and modifications in export management policies will cause major changes for water users and managers throughout California. In this chapter, we estimate the costs of different approaches to managing Delta exports and outflows from the perspectives of both water supply and quality. Although there is substantial expertise and knowledge of these costs at the local and regional levels, this knowledge has not been well integrated. We provide an initial attempt to synthesize these costs from a statewide perspective. Our estimates are not exact, but they form a reasonable basis for drawing some broad conclusions about the economic implications of different export alternatives.

STATEWIDE ADAPTATIONS TO DELTA WATER MANAGEMENT

The reliability of the Delta as a water source is of great concern to water managers, particularly those whose agencies rely on direct diversions of

Delta water. At issue are both regulatory reliability (given continued concerns over the needs of fish species) and physical reliability (given the threats to the integrity of the levee system). In response, many users of exported water have made strides to reduce their dependence on the Delta in recent years. Urban water agencies have been developing interties—or connectors between aqueducts—to enable water sharing in the event of emergencies, such as a massive levee failure. Both urban and agricultural water agencies have developed underground storage (or "groundwater banking"), water-use efficiency, water markets, transfers and exchanges, wastewater reuse, and other activities. Indeed, much can be done to reduce the dependence of water users on Delta supplies, although such actions always come at some cost, in terms of financial expense or water scarcity (i.e., using less than the desired amount of water).

If water supplies from the Delta were abruptly cut off and water users were both unable to draw on alternative supplies and unprepared to reduce water use, the results would be catastrophic for many users. Costs to water users for such scenarios, arising from multiple levee failures, are estimated to be as high as $10 billion per year (Illingworth et al. 2005). In contrast to these scenarios, this chapter examines a "soft landing" approach to adaptation, where reasonable preparations would be made for any major changes in Delta conditions and management.

Water suppliers and users can be remarkably adaptable, particularly over long periods of time. Studies of how California's water supply could adapt to major climate, population, and infrastructure changes indicate that considerable adjustment is physically possible at reasonable cost (Medellin-Azuara et al. 2008a; Tanaka et al. 2006; Jenkins et al. 2004). Adaptations and innovations are facilitated by California's highly intertied water system and decentralized water management. State and federal agencies manage the largest water projects, but many planning decisions are made at the local and regional levels. Local and regional water agencies commonly have the political, financial, and technological wherewithal to make long-term changes in their water supplies and water use. Although institutional conflicts often limit short-term actions, cooperation has increased considerably in recent decades in such areas as water marketing, groundwater banking, and emergency sharing agreements.

Table 7.1 summarizes many of the options available to water managers seeking to balance supplies and demands. In addition to traditional methods

TABLE 7.1 *Water Supply System Management Options*

Demand and Allocation Options

- General
  - Pricing[a]
  - Subsidies, taxes
  - Regulations (water management, water quality, contract authority, rationing, etc.)
  - Water transfers and exchanges (within or between regions/sectors)[a]
  - Insurance (drought insurance)
- Demand Sector Options
  - Urban water-use efficiency[a]
  - Urban water scarcity[a] (water use below desired quantities)
  - Agricultural water-use efficiency[a]
  - Agricultural water scarcity[a]
  - Ecosystem restoration/improvements (dedicated flow and nonflow options)
  - Ecosystem water-use effectiveness
  - Environmental water scarcity
  - Recreation water-use efficiency
  - Recreation improvements
  - Recreation water scarcity

Supply Management Options

- Operations Options (Water Quantity or Quality)
  - Surface water storage facilities (new or expanded)[a]
  - Conveyance facilities (new or expanded)[a]
  - Conveyance and distribution facility operations[a]
  - Cooperative operation of surface facilities[a]
  - Conjunctive use of surface and ground waters[a]
  - Groundwater storage, recharge, and pumping facilities[a]
- Supply Expansion Options (Water Quantity or Quality)
  - Supply expansions through operations options (reduced losses and spills)
  - Agricultural drainage management
  - Urban water reuse (treated)[a]
  - Water treatment (surface water, groundwater, seawater, brackish water, contaminated water)[a]
  - Desalting (brackish and seawater)[a]
  - Urban runoff/stormwater collection and reuse (in some areas)

[a]Options represented in the CALVIN model.

to expand usable water supplies, such as surface storage, conveyance, and water treatment, the list includes more contemporary methods, such as improvements in operational efficiencies and wastewater reuse. Water demand management measures include improving water-use efficiency ("more crop per drop"), as well as reducing water use beyond desired levels by rationing urban water use, fallowing some farmland, or curtailing recreational activities. Various general tools (pricing, water markets, exchanges, and taxes or subsidies) may be used to motivate local users to implement both supply- and demand-side options.

A wide range of alternatives exist for managing Delta water supplies. As seen in Chapter 2, numerous alternatives have been proposed in the past. Various Delta outflow regulations, policies on Delta exports, changes in physical pumping, conveyance, and storage capacities would be reasonable elements to examine, both individually and in combinations. If one also considers a reasonable set of adaptations by water users and managers, estimating the performance of alternatives becomes a complex exercise. Here, we draw on results of modern computer models to examine the ability of California water users to adapt to changes in water supply available from the Delta. The CALVIN (California Value Integrated Network) model explores how California's larger water supply system could respond to changes in water supplies and demands resulting from different Delta management strategies. Other studies have been done on in-Delta agricultural impacts of water management (Lund et al. 2007).

All model results are based on imperfectible assumptions and limited information. Nevertheless, for such complex systems as the Delta and California's water supply, such analytical aids are indispensable for exploring, developing, and evaluating new alternatives. Computer models allow us to represent current knowledge and explore the implications of uncertainties in a standardized evaluation of a wide range of solution alternatives. Although there are obvious pitfalls to quantitatively analyzing such complex systems, making decisions without such aids has shown itself to be risky, even dangerous. Model results provide insights based on our best knowledge of the system and are a relatively transparent way to compare policy and management alternatives for complex systems.

COSTS OF PROVIDING MORE WATER FOR THE ENVIRONMENT

Under the present through-Delta pumping system, water exports from the Delta raise two major environmental concerns: (1) entrainment of fish and disruptions of fish movement by the export pumps in the southern Delta, and (2) the volume and timing of net fresh water outflows from the Delta to the sea, which affect the location, extent, and variability of habitats available to various species through the course of their life stages (see Chapter 6). Both issues are affected by the quantities of water exported from the Delta, as well as by a host of other aspects of internal Delta water management and water export characteristics, such as location of exports, operating pattern, and specific design of facilities. Net outflows from the Delta are also affected by the volume of upstream diversions, which are nearly twice as large as volumes exported from within the Delta (see Chapter 4).

For several decades, exports have been regulated in various ways to protect fish and Delta agriculture and urban uses, most notably with minimum flow requirements and maximum salinity standards at particular times of the year. Judge Wanger's ruling in late 2007 has led to further restrictions on export pumping to reduce the risk of entrainment of delta smelt (see Chapter 2). Other recent discussions suggest the potential for additional regulatory actions. In light of fish population declines, both the Delta Vision Blue Ribbon Task Force (Isenberg et al. 2008a, 2008b) and many environmental advocates have argued for considering a future with reduced exports, with export users relying more on local supplies and conservation. The task force also indicated that upstream water users should contribute by limiting their use of the waters flowing into the Delta.

A major policy question is: How will potential reductions in export levels and upstream diversions affect individual water users and the wider economy? Water users have many ways to adjust to cutbacks, each of which entails some costs. Water users throughout California's main population centers and farming regions are tied to an extensive water storage and conveyance system, including groundwater and surface water storage, canals, pipelines, pumps, hydropower turbines, and water and wastewater treatment plants (Figure 7.1). Local supplies can also be expanded through treatment of wastewater, construction of desalination facilities, and new conveyance and storage. Furthermore, water users can manage their own water demands

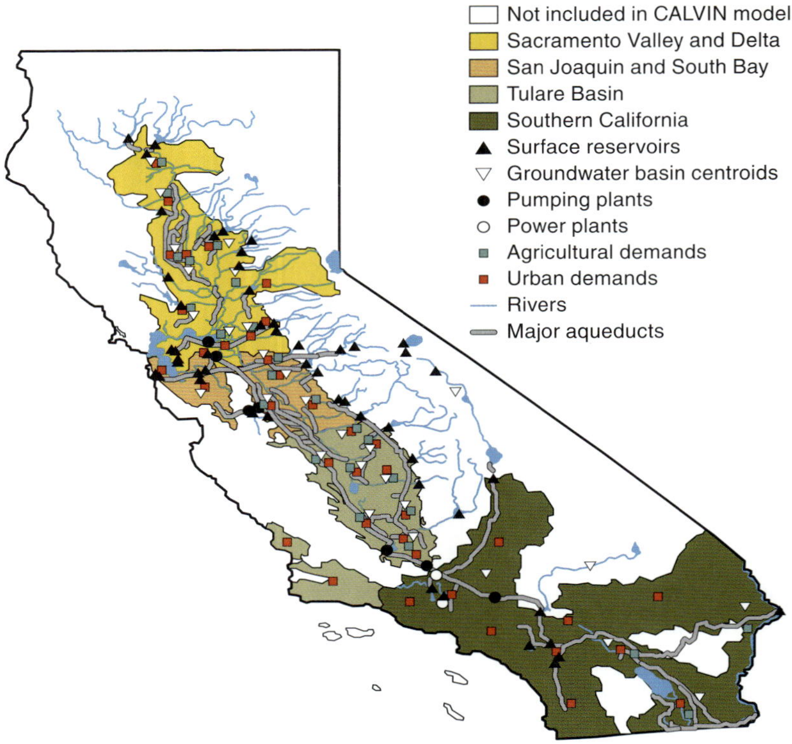

FIGURE 7.1 California's statewide water supply network. The figure shows the water system represented in the CALVIN model, discussed in the text. Areas shown in white have localized water systems, not highly connected to the statewide system.

(through conservation and rationing) or buy water from others who have water uses of lower value. In short, water users have considerable ability to adapt to changes in how the Delta is operated. Some adaptations are likely to be more costly than others, presenting higher operating costs or imposing greater water scarcity (or shortage)—lost profit for farmers and greater expenses and inconvenience for urban users.

To take into account the many options for adapting to changes in water availability, it is necessary to use a computer model of the California water system. Here, we used the (California Value Integrated Network) CALVIN model of California's statewide water supply system.[1] This model suggests economically promising portfolios of water management

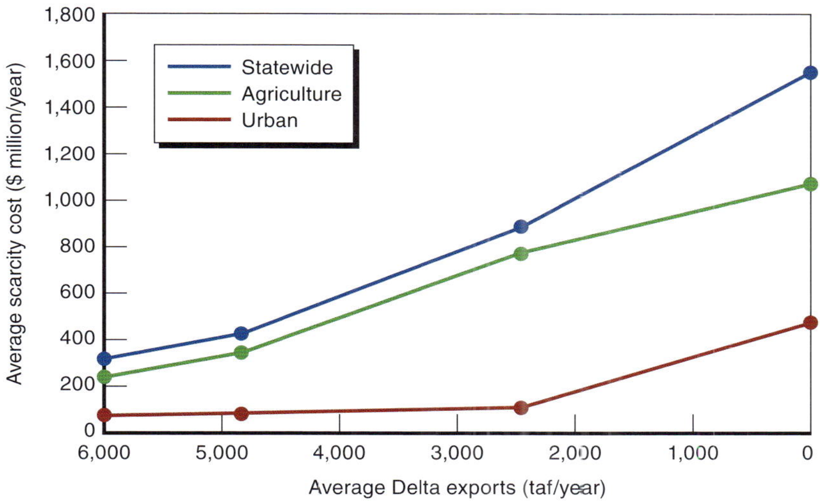

FIGURE 7.2 Annual average statewide scarcity costs with changing export restrictions, 2050. Source: Tanaka et al. 2008.

activities in response to a set of economic, population, climate, policy, and infrastructure conditions. Because we are interested in assessing how water users would adapt to long-term changes in Delta policy, we examined scenarios with population and land use conditions at the middle of this century (2050). Details of the results appear in Tanaka et al. (2008).

REDUCING OR ENDING WATER EXPORTS

Figure 7.2 depicts the statewide costs of water scarcity (or shortages) from a planned reduction in water export volumes, starting from a 2050 baseline demand of approximately six million acre-feet and declining to no exports whatsoever. Even at this baseline level, water users experience some water scarcity costs—on the order of $300 million per year statewide. Costs of initial cutbacks are relatively small, but they rise significantly for the agricultural sector once exports are reduced by more than one million acre-feet.[2] The urban sector begins to experience significant scarcity only when exports are restricted to less than half their initial volume. Cities would avoid the full brunt of cutbacks by purchasing water from southern Central Valley farmers who currently use local inflows and employ more wastewater reuse, seawater desalination, and water conservation. Although cities would face higher water prices as a result of these shifts, southern Central Valley agriculture would see the greatest overall effects from the cutbacks, from the loss of water supply and, for some, from selling available water to Southern California urban users. Individual farmers who were able to sell water could experience financial gains, but local communities would experience a loss in economic activity related to farming.

When additional net operating costs of roughly $200 million are added to these water scarcity cost estimates, a planned ending of Delta exports is estimated to directly cost the statewide economy roughly $1.5 billion per year (2008 dollars). Because this cost estimate is based on modeling that assumes that water managers have perfect knowledge about future hydrologic conditions and face no institutional constraints to water marketing, the real costs would likely be higher. For instance, if farmers elected not to sell more water to urban areas once exports were ended, the total cost to the economy would jump to $2.2 billion, as urban water users adopted higher-cost sources. Allowing for other inefficiencies and delays, the upper bound on statewide costs of ending exports might be as high as $2.5 billion.

Small reductions in exports are significantly less costly because there is the possibility of reallocating water from lower-valued (mostly agricultural) uses, through the water market (including transfers from lower-value to higher-value farming—as often occurs today; Hanak 2003). Even for large reductions or elimination of water exports, the economic costs, although large, are not catastrophic for California's $1.8 trillion per year statewide economy. Planning for such a transition significantly lowers costs. By comparison, an unplanned, temporary interruption of exports from a catastrophic failure of Delta levees is estimated to cost water users from $8 to $16 billion (Illingworth et al. 2005; URS Corporation and Jack R. Benjamin and Associates 2007b). Other studies confirm the high costs of unplanned reductions in exports arising from recent regulatory or drought-related cutbacks.[3]

Seen from another perspective, our estimate of total economic costs of ending Delta exports can mask the social consequences for specific regions. In particular, this estimate measures losses to the southern Central Valley agricultural sector in terms of forgone returns to land and farm management when land is taken out of production—a loss of roughly

$800 million. But the cutbacks in water (a decrease of 29 percent) and acreage (877,000 acres, or 26 percent) imply a substantially greater loss in regional revenues and employment. Farm revenues would drop by $3.3 billion (a decrease of 17 percent), regional revenues by $4.4 billion, and regional employment by between 100,000 and 130,000, including about 40,000 agricultural jobs (Howitt et al. 2009). These results highlight the disproportional effect of ending exports on the southern Central Valley.

INCREASING NET DELTA OUTFLOWS

Under the current system, where exports are drawn through Delta channels to the pumps, directly reducing exports could avoid or lessen entrainment and other problems created by altered flows within the Delta. If, instead, exports are diverted around the Delta through a peripheral canal, the pumps no longer play a direct role in the Delta, and the regulatory issue is mainly one of maintaining appropriate flows into and out of the Delta. Increased net Delta outflows also could be sought to maintain salinity standards for agricultural and urban users within the Delta in the face of sea-level rise, which is likely to draw sea salts from San Francisco Bay further into the Delta (see Chapter 5).

Restrictions on water exports can increase net Delta outflows by reducing the amount of water diverted from the system. However, the goal of increasing outflows can be attained more directly (and cost-effectively) by regulations requiring increased outflows (Figure 7.3). Even if export users continue to have the regulatory responsibility to ensure that such flow requirements are met, outflow requirements allow more senior upstream diverters to participate in the solution by leasing or selling some of their water to export users. Figure 7.3 compares the costs of increasing Delta outflows for these two regulatory approaches. The red dashed curve shows the costs of using reduced export requirements. At the left-hand end of the curve, annual export levels are roughly six million acre-feet (maf) and average net Delta outflows are roughly 13 maf. When the restrictions reach their maximum level, with zero exports (the right-hand end of the "reduced export requirement" curve), average net Delta outflows are 18.7 maf per year. It would be possible to achieve the same volume of outflows with a direct outflow requirement (the blue curve) at a significantly lower cost to the economy: $1.1 billion per year lower.

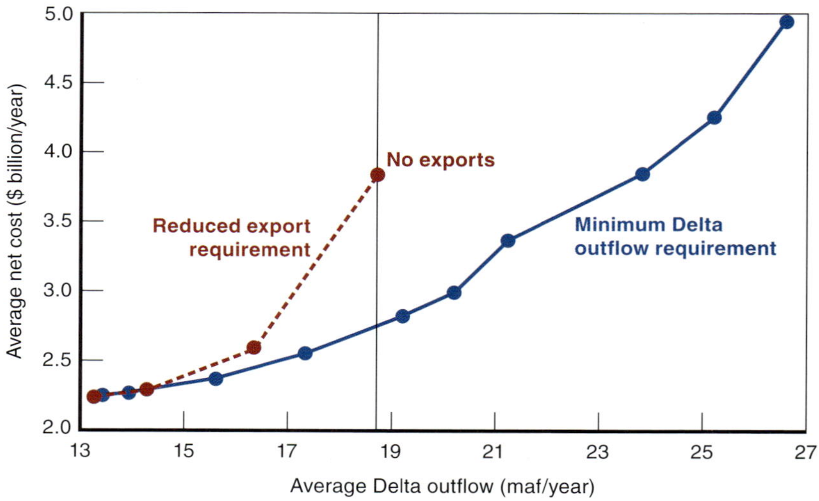

FIGURE 7.3 Statewide costs of alternative regulations for Delta outflows, 2050. The reduced export requirement curve shows the average Delta outflows with exports ranging from recent levels (around 6 maf/year) to zero. Source: Tanaka et al. 2008.

THE VALUE OF NEW INFRASTRUCTURE

We also used the CALVIN model to estimate the economic value of expanding water management facilities, such as surface storage and conveyance, under these regulatory alternatives. If water exports were substantially restricted or ended, it would be quite valuable to expand some local and regional conveyance facilities and add more interconnections among existing facilities. Such investments would increase the capacity of water users to transfer water and to benefit from regional investments in new supplies. In contrast, additional water storage rarely looks economically promising under these reduced export scenarios. Exceptions include additional groundwater recharge in the Bay Area and Southern California and local surface storage in Southern California (Tanaka et al. 2008). Similar patterns emerge when net Delta outflow requirements are increased.

URBAN AND AGRICULTURAL WATER QUALITY

The increasing salinity of Delta water with sea-level rise and island flooding raises questions about the economic costs of different export management alternatives from various quality perspectives. Here we summarize estimates of the effects of Delta water quality on drinking water treatment costs—an important consideration for the many urban users of Delta waters—and on agricultural revenues in the southern Central Valley.

DRINKING WATER TREATMENT COSTS

Salts and organic compounds present in Delta waters increase drinking water treatment costs and health risks for urban users in the Bay Area and Southern California. Under today's conditions, the Delta already has significantly lower quality water than upstream locations on the Sacramento River, where intakes to a peripheral canal would be located. We estimate that the costs of treating drinking water from the Delta are currently $20 to $60 per acre-foot higher than if water were taken from the Sacramento River upstream of the Delta. With sea-level rise or western island flooding, the additional cost of using the Delta as a source of urban water supply would rise to $100 to $500 per acre-foot (Chen et al. 2008). The deteriorating quality of Delta water also raises cancer risks from ingestion of disinfectant by-products, which are not normally removed during treatment.

AGRICULTURAL LOSSES

Until recently, 3.7 maf per year of water were exported on average from the Delta to agricultural water users in the southern Central Valley (Table 4.1). These exports also delivered 1.5 million tons of salt (at 300 mg/l) per year, resulting in approximately 500,000 tons of net salt accumulation in these agricultural lands (Orlob 1991; Schoups 2004). The accumulation of salts is steadily degrading the productivity of agriculture in parts of the southern Central Valley, particularly on the western side. Salinity lowers crop yields, prevents the farming of some higher-value crops, and can ultimately render land unprofitable for agriculture. Salinity also constrains confined animal feeding operations, because it limits where animal wastes can be applied.[4]

If this region continues to experience recent levels of water exports at current levels of Delta salinity (3.7 maf/year at 300 mg/l) and farmland salinization continues at past rates, additional losses of agricultural revenues resulting from salinity could reach $392 million per year by 2030 (Figure 7.4).

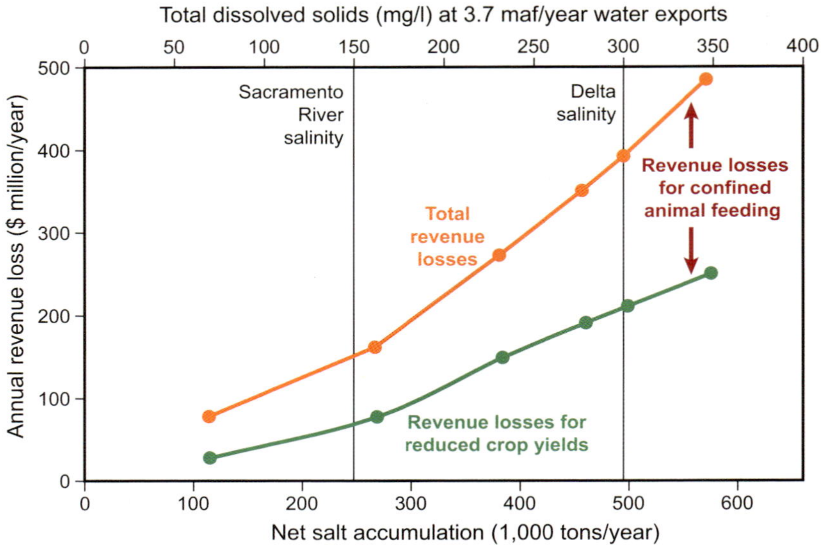

FIGURE 7.4 Agricultural revenue losses from annual salt accumulation in the southern Central Valley, 2030. Source: Medellín-Azuara et al. 2008b.

This salinity cost will average about $105 per acre-foot of water delivered, with a marginal cost of roughly $135 per acre-foot.[5] Reducing export salinity to the levels in the Sacramento River (150 mg/l) might lower these costs by as much as $241 million per year (Figure 7.4) or $65 per acre-foot of water delivered. These water quality benefits will be lower if current trends in the growth of shallow saline areas do not continue or if reduced salt loads do not slow the growth of salt-affected areas. Taking into account these factors, salinity costs to agriculture in 2030 from using Delta water rather than Sacramento River water appear to lie in the range of $210 million to $270 million per year, with a resulting loss of 5,000 jobs in agriculture and 16,000 jobs overall from these water quality effects alone (Howitt et al. 2009). Although these estimates are preliminary, they suggest that as salts accumulate, the salinity costs to southern Central Valley agriculture are substantial and will grow with time. Indeed, these estimates may be conservative insofar as they assume continuation of exports at current Delta salinity levels, not the increased levels that are likely with sea-level rise (Figure 5.7).

IMPLICATIONS FOR EXPORT MANAGEMENT ALTERNATIVES

The likely ranges of export water quality and quantity vary with each of the four water export strategies, as do their economic costs and benefits.

CONTINUED THROUGH-DELTA PUMPING

With continued reliance on through-Delta pumping to meet export water demands, export water users would face continued supply uncertainties and deteriorating water quality. Exports are likely to become increasingly unreliable, with reduced average quantity and quality because of sea-level rise, island failures, and uncertainties in the effects of exports on endangered species. The high costs of treating Delta water for drinking would continue or increase, as would the import of salts into the southern Central Valley, hastening the end of farming in many areas. Eventually, with sea-level rise and island flooding, increasing salinities would reduce water exports in all but the wettest years and Delta exports would diminish and likely eventually end.

PERIPHERAL CANAL

Exports taken from the lower Sacramento River would have higher water quality for urban and agricultural purposes and lower water quality costs. If all exports were taken upstream of the Delta, with recent (pre-Wanger decision) levels of exports, the water quality benefits alone could be $300 million to $1 billion per year by the middle of this century. Whether a peripheral canal would entail economic costs from reductions in export deliveries would depend on the operational and environmental aspects of water management. Upstream intakes would avoid most entrainment issues affecting Delta species. But to avoid entrainment of Chinook salmon and other species living in or passing through the Sacramento River and northern parts of the Delta, the canal might be required at times to divert less water than has occurred in the recent past (see Chapter 5), increasing costs to water users. Some overall reductions in exports also might be needed for overall environmental flow management in the Delta.

DUAL CONVEYANCE

Initially, a combination of canal and through-Delta pumping should provide greater water supply reliability than upstream diversion alone. In particular, if dual conveyance could capitalize on its potential physical flexibility to avoid fish entrainment and related problems, it might face fewer regulatory restrictions on exports than either through-Delta pumping or a peripheral canal alone. However, dual conveyance offers more limited water quality benefits than a peripheral canal because of the higher salinity of Delta waters. If all urban water was taken from a peripheral canal and all agricultural water continued to be channeled through the Delta, the net water quality benefits as compared with a through-Delta strategy alone would still be roughly $100 million per year today. Such water quality segregation would require more complex operation of the California Aqueduct (as batch pipelines from oil refineries do now) and perhaps additional near-Delta storage. Over the long run, the water supply reliability of dual conveyance would diminish as sea-level rise and island failures curtail the use of the through-Delta component.

NO EXPORTS

Ending water exports as a long-term water supply solution would probably cost at least $1.5 billion per year, or perhaps as much as $2.5 billion—a substantial sum, but not catastrophic for the statewide economy. However, ending or severely reducing exports would be catastrophic for many agricultural areas in the southern Central Valley. This strategy would also reduce the economic basis for funding extensive environmental investments in the Delta.

In sum, from the perspective of the statewide economy, there are clear advantages to moving toward a peripheral canal or dual conveyance system. In addition to reducing the risks of costly disruptions in the water supply from a failure of Delta levees, these options have the potential to reduce the regulatory costs for export users relative to the current through-Delta pumping system. They also provide a substantial windfall in water quality savings for both urban and agricultural water users. However, the benefits and costs of these alternatives might not be equally distributed, depending on the governance and finance policies implemented.

COSTS FROM UNREPAIRED DELTA ISLANDS

Although not repairing some Delta islands after their levees fail makes economic and business sense, it does imply losses of agricultural production and costs to the Delta's regional economy. These costs amount to about $93 million per year in reduced agricultural profits, $124 million per year in reduced regional income, and 2,800 jobs lost from the Delta's regional economy for 19 permanently flooded subsided islands retired from agriculture.[6] This is significant, but a small proportion of the economy of the five Delta counties. These costs and losses are inevitable long-term consequences for the Delta economy as these islands fail, regardless of which water export strategy is chosen. Although a peripheral canal might entail losses to some Delta agriculture as a result of increased salinity, these losses are likely to be much smaller in magnitude. Given the long-term risks of island flooding, however, there is no long-term trade-off between Delta jobs and agricultural jobs in the southern Central Valley.

Overall, the selection of a water management and operations strategy for the Delta will have important consequences for water costs, regional income, and jobs in many parts of California. While these costs are large, particularly at the local level, they are not necessarily catastrophic in the context of a $1.8 trillion state economy. Yet, regardless of the water export strategy chosen and how water exports are operated, in the long term some Delta islands will fail permanently, with some economic consequences for that region.

8

POLICY AND REGULATORY CHALLENGES

> If a man neglect to strengthen his dike and do not strengthen it, and a break be made in his dike and the water carry away the farmland, the man in whose dike the break has been made shall restore the grain which he has damaged.
>
> *The Code of Hammurabi* (circa 2250 BCE), translation by ROBERT FRANCIS HARPER (1904)

To increase the chances of favorable ecosystem and economic outcomes, California needs a policymaking environment that enables decision makers to anticipate the changes facing the Delta. This requires effective political leadership, a sound governance and finance system, and an appropriate set of regulatory tools.

Given the many often-conflicting stakeholders concerned with Delta outcomes, there is no substitute for higher-level political leadership to help chart a new course for Delta management and negotiate solutions for some of the difficult trade-offs among human users of the Delta's resources. Mitigation offers a promising path for resolving some of these trade-offs, while fostering policies in the best overall interests of the state. However, given long-term limitations on state and federal funding, it is in both state and local interests for beneficiaries to pay for most Delta actions, rather than delaying urgent decisions with the distracting notion that state and federal governments will provide most funding. To answer the thorny question of ensuring stable funding for ecosystem management, California will also need to move beyond the recent model of relying on periodic injections of state bond funding.

One central issue for Delta governance is setting up better oversight of regional land resources, given the pressures for urbanization and the public safety and environmental risks this poses from floods and earthquakes.

Although some attempts have been made to provide oversight, land use regulation in the Delta is less effective than in other environmentally sensitive areas in the state. Another key governance issue brings California into new territory: how to provide adequate environmental and political safeguards in the event that a peripheral canal or dual conveyance system is built.

There are also questions about whether the current regulatory framework is compatible with the changes coming in the Delta, either as a result of human actions (such as a peripheral canal) or of natural forces (notably, climate change). First, does the current federal and state system for managing Delta water quality allow for anticipatory, versus reactive, interventions? Second—as suggested by the quotation at the beginning of this chapter—what does the prospect of more Delta levee failures and island flooding mean for local and state responsibilities to neighboring landowners? Third, how can upstream diverters become part of a Delta solution? And fourth, how are Delta solutions that aim to balance ecosystem and economic goals likely to fare in the face of an increasingly difficult natural environment for desirable species?

In this chapter, we provide an overview of financing and mitigation options and models for regional land management. We then focus on these four regulatory questions and the governance issue of providing safeguards for a new Delta.[1] Our intent is not to provide the final word on these issues but rather to highlight areas that will need to be addressed squarely as part of any long-term Delta solution.

FUNDING PRINCIPLES FOR A SOFT LANDING

By our rough estimates, the water supply infrastructure costs of a Delta solution are likely to be in the range of $5 to $10 billion dollars, with significant additional sums required for ecosystem restoration efforts, stronger urban levees, and adjustments by other infrastructure providers (see Chapter 9 and the appendix). It is not realistic to expect taxpayer dollars to meet all, or even most, of these costs, given other demands on public spending. Beyond this, general obligation bond financing of water supply infrastructure (repaid with general state tax revenues) establishes poor incentives for local water managers to operate efficiently. If someone else is paying, it is always easier to ask for more. Thus, it is necessary to consider other options.

THE BENEFICIARY PAYS PRINCIPLE IS MORE RELEVANT THAN EVER

One of CALFED's clear failures was its inability to mobilize adequate stakeholder contributions to its investment portfolio (Little Hoover Commission 2005). Everyone had signed on to the "beneficiary pays" principle, but stakeholders tended to argue that program elements provided public benefits, and should therefore be funded by state and federal taxpayers (Misczynski 2009). General obligation bonds were used to support investments in groundwater storage, water recycling, and water-use efficiency—making it cheaper for water agencies to stretch or expand water supplies—on grounds that this lessened pressures on the Delta ecosystem. Water users who stood to gain from new surface storage investments made similar arguments, without offering to fund a significant share themselves. Today, some Delta stakeholders are calling for massive public investments in Delta levees, even though many of the beneficiaries are clearly private or localized in nature: water export users, Delta farmers and land developers, power and rail companies, and users of the local road network.

Because the costs of any new Delta strategy are likely to greatly exceed the funds available from state and federal coffers, better ground rules on financial contributions are needed. User finance—that is, payment by the actual users of the investments—has many advantages. It frees public funds for truly public purposes, such as ecosystem restoration and mitigation, and it helps ensure that many investments are cost-effective. If water users are unwilling to finance investments that increase the reliability of their water supply, chances are that the investment is not a sound one. If landowners are unwilling to contribute to the costs of flood protection, chances are that the value of the land to be protected is too low to merit such investments.[2]

User contributions would be especially relevant for collective infrastructure investments in both water supply and flood protection. Water export users should be expected to fund improvements in water supply reliability, and a variety of beneficiaries should be expected to contribute to programs to reduce flood risks. It is often argued that mobilizing user contributions to Delta flood control is too complex, given the many interests involved and the fact that some of them—such as Caltrans—lack specific budgets to pay for such programs. But straightforward precedents exist for user finance in other areas of public safety. For example, the private sector finances most investments in seismic retrofits and prevention; Bay Area bridge users pay a surcharge to help fund seismic retrofits of bridges. There is no reason why a beneficiary pays principle could not apply to infrastructure adjustments in the Delta. For new homes and businesses, developer fees are a straightforward way to collect up-front contributions to flood protection, and property assessments can be used to cover maintenance costs. The key challenge is to ensure that these fees and assessments are high enough to cover the costs of building and maintaining adequate protection. If not, the local community (and state taxpayers) will be left footing the bill.

APPORTIONING COSTS OF LARGE WATER PROJECTS

For water supply investments large enough to require the participation of multiple parties, one stumbling block facing CALFED was the lack of agreement on how to apportion costs among beneficiaries: Should each water user be required to pay the same amount for each unit of water received, or might some sort of sliding scale be appropriate? This question is particularly relevant for Delta exports. In a typical year, agricultural water users have employed most (72 percent) of the direct diversions from the Delta (Table 4.1), yet most agricultural uses cannot justify costs as high as those urban users are willing to pay.

Two central problems facing any public project are how large to build it and how much to charge users to cover the costs of the project. Standard economic calculations of marginal cost pricing, whereby all users are charged the incremental project cost, typically will fail to recoup total costs of water projects because the incremental cost falls as the size of the project expands. These economies of scale occur because building water projects often involves a large fixed cost and a relatively small constant per unit operating cost.

An analogy can be made with the cost of operating a passenger jet. The costs of operating the jet are essentially the same regardless of the number of passengers. The incremental cost of a student in the back of the plane is little more than peanuts (the in-flight snack), so how much of the fixed cost of flying the plane should be charged to the student and how much should be charged to a business class passenger? One answer comes from the economist Frank Ramsey. Ramsey (1927) worked out that each user should cover the incremental costs (the peanuts) and that the fixed costs should be allocated in proportion to each user's price sensitivity—or the extent to which the quantity purchased varies with price. This rule is generally termed Ramsey pricing. Where such economies of scale exist, the Ramsey rule says that the least price sensitive group (business class) should pay the greatest proportion of the fixed costs through higher fares, and the most price sensitive group (the student traveler) should be charged the lowest proportion.

For water projects, users have wide variation in sensitivity to water prices—what economists refer to as the price elasticity of demand. Several studies have estimated the elasticity of demand for urban water users to be between −0.2 to −0.4 (i.e., a 10 percent increase in price, decreases urban use by 2 to 4 percent). Irrigated agriculture is more price responsive, with elasticities of demand for water ranging from −0.8 to −1.2 (implying a drop in use of 8 to 12 percent for a comparable 10 percent price increase). It follows that the practice, adopted by many water projects, of charging urban users higher prices than agricultural users, can be justified as efficient, permitting the overall service area to benefit from scale economies.

Similarly, urban water suppliers often charge commercial users more than residential users. In times of drought, those paying higher prices also are often provided with greater reliability (another economically efficient outcome). Such pricing principles are also common in the rail, electricity, and airline industries.[3] They would be appropriate for a peripheral canal as well as new, near-Delta surface storage investments that might complement new conveyance.

The key point is that if the beneficiary pays principle is to be implemented to cover all the costs of building a project, the size of the project must be balanced against different users' willingness to pay for different amounts of water. Ramsey pricing is one way to balance these issues. It provides a standard method for efficiently allocating costs that users are willing to pay. Public statements about having users pay are not effective if the project design does not account for their observed willingness to pay.

Project plans must also be backed by formal, up-front financial commitments. The State Water Project (SWP) and many local water projects provide sound precedents for the principle that water users should pay for the water infrastructure from which they will benefit. Delta solutions should be no different.

MITIGATING ENVIRONMENTAL DAMAGE

It is also appropriate to create programs of environmental mitigation for stakeholders who benefit from the chosen alternative when those benefits put pressures on environmental resources—including water users, land users, and the beneficiaries of the shipping channels that run through the Delta.

For export water users, mitigation is already a premise of existing Delta agreements, although there are debates over the appropriate level of support. They have been expected to participate financially in CALFED ecosystem restoration projects, and mitigation is an integral part of the Bay Delta Conservation Plan now under development (see Chapter 2 and the discussion that follows). As discussed later in this chapter, regulatory options also exist to incorporate upstream diverters into a solution. Rather than relying on public funds, a more appropriate—if more politically difficult—solution is to charge an ecosystem fee for all water diverted from the Delta. Tapping into the windfall savings in water quality would be a natural source of funds if a peripheral canal or dual facility were adopted (see Chapter 7).

Environmental mitigation should be required for the urbanization of Delta lands, given the irreversible changes caused by land development. One possibility would be to set aside requirements to maintain some lands for environmental uses. Such mitigations are already a standard practice for new development in many parts of the state. The Delta, with its unique environmental resources, should be no exception.

Environmental mitigation is also appropriate for ships using the Ports of Stockton and Sacramento, given the role of ballast water in introducing alien species. Present ballast water control requirements are too lenient to be of much value for the Delta. A ballast water fee could be imposed on shippers who do not undertake significant additional efforts. Tighter controls are also appropriate for horticultural, aquarium, bait, and other industries that deal with live organisms, all of which are likely sources of invasive species.

PUBLIC SECTOR FUNDING ROLES

Even with application of the beneficiary pays principle to collective investments in water supply, flood control, and environmental mitigation, public funds will be needed to implement a more sustainable long-term solution for the Delta. State and federal taxpayer contributions are appropriate to help finance programs for which the general public is a beneficiary, including some environmental restoration. In some cases, public benefits would include avoiding future public liabilities—a justification for taxpayer contributions to flood control and other emergency-preparedness measures. Public funds are also appropriate for programs considered important from the perspectives of equity and social justice—for example, programs to provide safe drinking water to low-income rural communities. Finally, public funds can provide incentives to encourage various stakeholders to agree to actions that would generate overall social benefits that they might otherwise be reluctant to pursue. These last two reasons justify using bond monies or other public resources to finance programs to soften the costs of adjusting to Delta solutions.

SOFTENING THE COSTS OF ADJUSTMENT

No matter which Delta alternative is chosen, all users of Delta services will face some additional costs. In all cases, water export users will need to make new investments to improve reliability and quality; under some alternatives, they would bear added water scarcity costs as well. Under any plan, some Delta farmers will go out of production because of island flooding; others will incur additional costs under regimes that feature fluctuating Delta salinity, notably with a peripheral canal–only alternative. Sooner or later, urban water users that pump directly from the Delta will need to alter their intake points, possibly building aqueducts to connect to more reliable freshwater sources.

The increasing flood risks that accompany climate warming and sea-level rise will also carry adjustment costs. Existing and planned urban areas behind Delta levees will need to invest in levee upgrades. The owners and users of the various types of infrastructure that crisscross the Delta will face additional costs for these same reasons. Suisun Marsh duck clubs will find it increasingly difficult to keep salt water from breaching their fragile levees and will eventually need to shut down or move elsewhere. And although recreational boating will continue in any likely future, alternatives that modify the channel network could reduce revenues at some local harbors. Recreational fishing, while likely to increase with the expansion of aquatic habitat, will also face some costs to accommodate such expansions.

CANDIDATES FOR MITIGATION

Clearly, it is neither feasible nor desirable for state taxpayers to compensate all of these interests; doing business in the Delta is becoming more expensive because the current system is unsustainable, not because of the actions of the state or any one group. However, mitigation can soften the costs of adjustment for interests that will be particularly hard hit by changes to the status quo. For management alternatives that move away from a freshwater Delta, this list includes Delta farmers and urban agencies that draw water directly from the Delta (notably, Contra Costa Water District). For alternatives that also significantly reduce water exports, this list includes farmers on the west side of the San Joaquin Valley and in the Tulare Basin. For alternatives that result in significant water transfers, this list might also include communities in the source regions. Other candidates could include owners of land that would benefit environmental goals—e.g., the Suisun Marsh duck clubs—or businesses that would be affected by changes in Delta channels.

There are no hard and fast rules for drawing up such a list. The goal of a mitigation process should be to encourage buy-in from interests that are likely to resist changes that could benefit the system as a whole. One consideration is legal standing. Under current agreements, Delta farmers and urban pumpers have protections on water quality (salinity) standards to the extent that these are affected by Central Valley Project (CVP) and SWP exports, as discussed later. Another consideration is equity. As Chapter 3 describes, some Delta land and business owners will face hardships because of island flooding, under any alternative. Likewise, some farmers in the San Joaquin Valley would lose substantially under the no-export alternative (see Chapter 7). It makes sense to consider mitigation options to help ease transitions in these communities, whether or not there is a legal obligation to do so.

MITIGATION OPTIONS

Mitigation does not imply a wholesale buy-out or coverage of all adjustment costs. Over time, the natural forces at work in the Delta will reduce the reliability of Delta services, requiring various groups to adjust anyway, largely at their own expense. Because almost all interests face worsening conditions, mitigations could be considered in relation to future "no action"

TABLE 8.1 *Potential Mitigations for a Delta Solution*

Mitigation Type	*Possible Targets*	*Precedents*
Investment cost sharing	Conveyance or storage investments for western Delta water users to allow introduction of more natural flow regime	State grant to Los Angeles for indoor conservation measures to reduce diversions from Mono Lake region
Financial compensation	Delta land purchases, either up-front or upon withdrawal from farming; subsidies for eco-friendly farming practices in Delta	State purchase of Sherman Island, federal purchase of some drainage-affected San Joaquin Valley farm lands; federal and local conservation easements
Physical substitution	Provide alternative lands for Suisun Marsh duck clubs	Relocation assistance is common with dam construction projects
Community mitigation funds	Support to communities affected by the loss of Delta lands from island flooding, increased salinity or loss of San Joaquin Valley farmlands from reduced exports	Some California water transfers; U.S. trade adjustment assistance
Performance bonds for environmental risk	Use to insure activities dependent on Delta water supplies (which risk cutbacks if fish do not recover)	Used commonly to cover risks of cost overruns and delays in large construction projects (environmental risk would be new)

SOURCE: Lund et al. 2007, Chapter 9.

conditions and effects, rather than in relation to some rosy, and unrealistic, continuation of current or past conditions. In other words, mitigations and compensations do not assure specific future performance or actions. Rather, they provide a *substitute* for assured performance.

Policies to soften adjustments could include a range of different forms of assistance, as summarized in Table 8.1. Many of these have been used in various contexts both in California and elsewhere.

BRINGING DELTA LAND USE INTO THE FOLD

As California's population growth has shifted inland over the last two decades, Delta farmlands have become a target for residential development. In the mid 2000s, the California State Reclamation Board estimated that as many as 130,000 new homes were being planned within the Delta. Although the recent economic downturn has brought new construction in the Central Valley to a virtual standstill, pressures to build in the Delta will resume when the economy improves. The increase in the number of homes along the perimeter and within the Delta will inevitably shift priorities for Delta management toward flood control and infrastructure to support urbanization. In addition to the concerns for public safety in this low-lying area, there are concerns that urbanization will come at the expense of habitat protection and other goals of statewide interest.

Thus, long-term solutions for the Delta need to address land use. Developing a governance framework that incorporates land use is particularly daunting in this region, given the current state of institutional fragmentation. Individual cities and counties are the permitting authorities for new development, and local reclamation districts are responsible for most decisions on levee maintenance and upgrades. There is little effective representation of larger regional and statewide interests in Delta land use decisions.

The Delta Protection Commission, established by the Delta Protection Act of 1992, is the only body representing regional interests in the Delta.[4] Its membership includes representatives from Delta cities, counties, and reclamation districts as well as various state agencies with Delta interests. Its primary purpose is to oversee land use and resource management issues in the Delta's primary zone, which the act reserved principally for agricultural, recreational, and environmental uses (Figure 8.1). Recently, the commission has begun serving as a regional forum for discussing growth issues more broadly. Although the commission may challenge land development that is inconsistent with the land use goals for the Delta's designated primary zone, it has no permitting authority and no ability to block land development.[5]

The Central Valley Flood Protection Board (formerly known as the State Reclamation Board) has the potential to exercise land use oversight in the Delta, through its authority to maintain the integrity of the flood control system. However, it has taken little interest in the Delta to date. Under

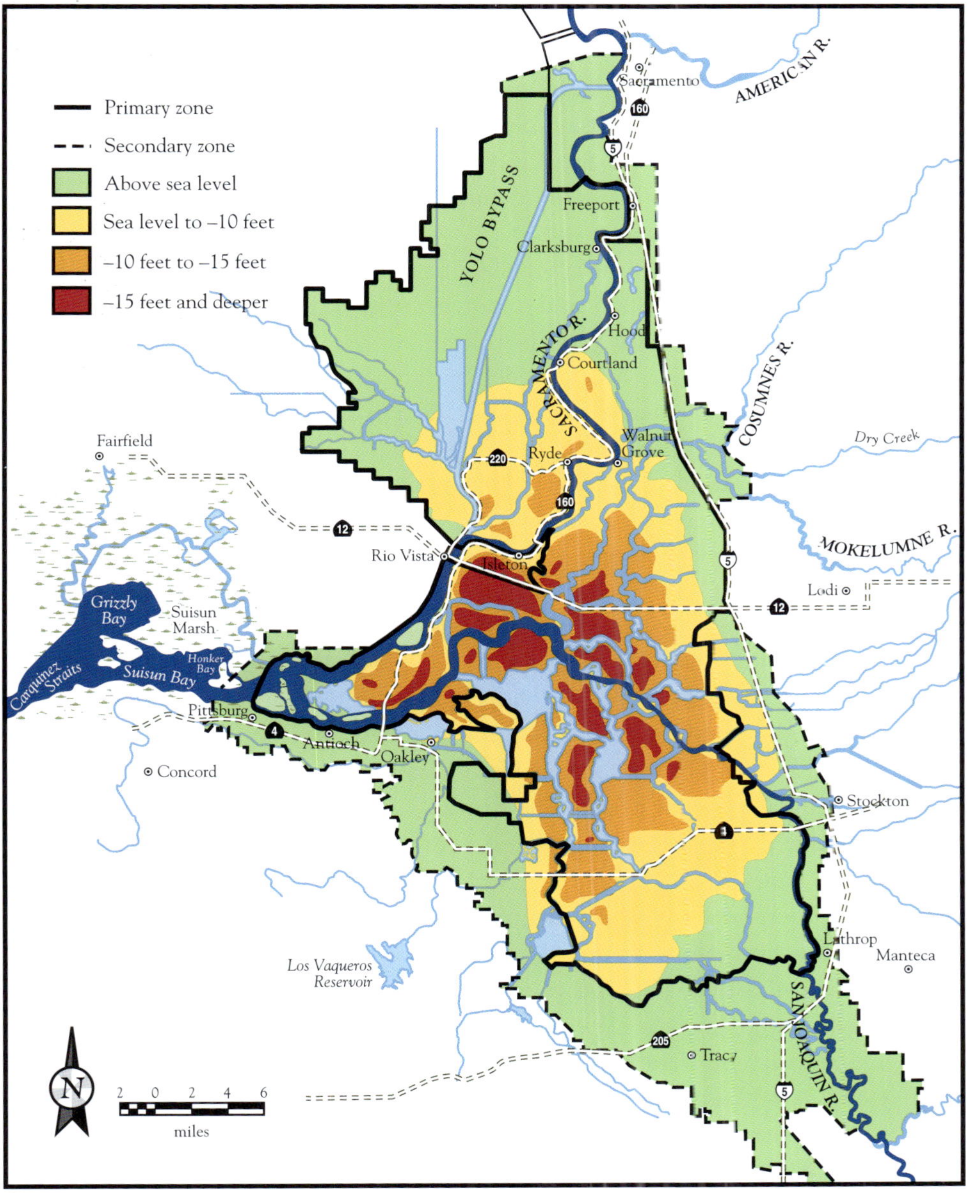

FIGURE 8.1 The Delta's primary and secondary zones.

current policies, it focuses only on issues that either directly affect project levees (just over a third of all Delta levees—see Figure 2.1) or increase regional flood levels. The board was at the center of a recent controversy over the River Islands housing development on Stewart Tract, with critics concerned that this decision did not adequately consider the implications for future flood risk either within the development itself or in neighboring areas (Lund et al. 2007).

The current lack of institutional authority for Delta land use, at a time when pressures to develop land resources are great, points to the need for a new approach. Although the issues are complex, there are many successful resource management models to draw on elsewhere in California, including regional authorities such as the San Francisco Bay Conservation and Development Commission (SFBCDC), the California Coastal Commission, and the Tahoe Regional Planning Agency (TRPA)(Aitchison 2007; Isenberg et al. 2008b).

The SFBCDC, the Coastal Commission, and the TRPA are permitting authorities that have successfully wrestled with balancing environmental and economic development goals in environmentally sensitive areas. These bodies were created in response to pressures similar to those that now face the Delta. The SFBCDC, in operation since 1965, was established to tackle problems of uncoordinated development that were leading to the filling of the San Francisco Bay, which lost an average of four square miles per year between 1850 and 1960. The Coastal Commission, established in 1972, was created to ensure that land and water uses in the coastal zone are environmentally sustainable. The TRPA was created in 1969 in response to two decades of rapid growth in the Lake Tahoe area, and charged with overseeing development. Unlike the Delta Protection Commission, these bodies have regulatory authority over a wide range of activities that have the potential to affect the beneficial uses of the land and aquatic resources in their jurisdictions. Both the SFBCDC and the Coastal Commission are authorized, under the Coastal Zone Management Act, to exercise regulatory oversight of the actions of federal agencies.

In addition to oversight of development is the question of managing environmental lands, to foster improved habitat conditions for the Delta's aquatic and terrestrial species. To manage these resources in a coordinated way, various forms of public or nonprofit entities may be appropriate. Here again, existing models can be useful, including a state land conservancy or a nonprofit land trust (Aitchison 2007; Isenberg et al. 2008b). A land trust model is particularly compatible with the continued private management of some lands for eco-friendly agriculture. Land trusts across California and the country have been important in developing conservation easement programs for farm and ranch lands. Some trusts play an active role in environmental land management as well.[6] One example within the Delta is the Cosumnes River Preserve, a 40,000 acre wildlife area managed by The Nature Conservancy and the U.S. Bureau of Land Management, in cooperation with other governmental and nonprofit partners.

The Delta Vision effort has emphasized the importance of these issues as part of a comprehensive solution for the Delta (Isenberg et al. 2008b; Natural Resources Agency 2009). For development oversight, it has suggested a hybrid model, which would strengthen the authority of the Delta Protection Commission and extend its jurisdiction to those lands within the Secondary Zone that are particularly at risk. A state conservancy has been proposed to manage environmental lands.

REGULATING WATER QUALITY IN A CHANGING DELTA

Since the CVP came on line in the 1940s, Delta water quality has been managed to keep salinity low enough for in-Delta agricultural and urban users and project beneficiaries south of the pumps. After the SWP became operational in the early 1970s, the two projects assumed joint legal responsibility for meeting specific water quality standards for in-Delta users. Over time, water quality standards have been added to protect fish species. The State Water Resources Control Board (the "Board") has primary authority for adopting water quality standards under federal and state law (respectively, the Clean Water Act, adopted in 1972, and the Porter-Cologne Act, adopted in 1969).

The Bay-Delta Water Quality Control Plan (WQCP) is the foundational document for Clean Water Act and Porter-Cologne compliance, and it includes measures to protect the legally designated beneficial uses of Delta waters: agriculture, municipal, and industrial uses, and fish and wildlife. The most recent WQCP, finalized in 1995 and updated in 2006, maintains preexisting standards for agricultural and urban diverters. To protect fish, it also includes a variety of minimum flow requirements, as well as maximum salinity standards at the western edge of the Delta at some times of the year (the so-called "X2" standard). D-1641, adopted in 1999, is the associated water rights decision that designates the SWP and CVP projects as responsible for meeting these water quality standards.

Under this system, all parties are assumed to benefit from lower salinity in the Delta, and the amount of water exported can be reduced and reservoirs operated to maintain standards for fish and in-Delta diverters. For several reasons, this system is likely to run into increasing difficulties. First, the modeling results shown in Chapter 5 confirm concerns raised by some in-Delta interests: At current sea level, a peripheral canal for water exports will make it more difficult to continue to meet salinity standards for some in-Delta diversions (Figure 5.8).

Second, the modeling illustrates that sea-level rise or island failures alone will generate similar or worse salinity effects for many users of Delta waters. Failure of some western Delta islands—increasingly likely with sea-level rise and other pressures on the levees—will constrain or eliminate through-Delta pumping and many in-Delta diversions. Even if the levees in the western Delta remain intact, one foot of sea-level rise, which is quite possible by the middle of this century, could generate frequent violations of salinity standards for agricultural users pumping in the western and central Delta under any export management alternative (Figure 5.8). Reducing exports or upstream diversions may help maintain Delta salinity standards under some scenarios, but this strategy will become increasingly costly.

These changes in the Delta raise two types of conflict relative to current water quality standards. First, a conflict could arise because one set of users (dependent on exported water) could maintain or even improve water quality with a different system of water management (a peripheral canal), but another set of users would be left with deteriorating Delta water quality. Second, a conflict could arise over inconsistencies in the water quality standards for different uses. If, as discussed in Chapter 6, it is better for desirable Delta fish species to allow greater variability in Delta salinity conditions across seasons and years, this would require standards that directly conflict with those designed to meet agricultural and urban needs.

The current regulatory system is not prepared to resolve such conflicts. In the extreme, the changes from sea-level rise or island failures imply that it would no longer be practical to maintain standards for some currently designated uses of the Delta. Yet, although the Clean Water Act does not guarantee specific levels of water quality to designated uses of Delta waters, it does not allow states to remove designated uses if they are already being served.[7] This restriction is tied to the assumption that direct human actions are the only sources of harm to water sources; the Clean Water Act did not foresee water quality changes because of climate change, such as salinity intrusion. Likewise, the act assumes that standards for different designated uses do not inherently conflict, as would be the case in a variable Delta. The question facing Californians is: Can flexible solutions to water quality conflicts be devised to allow proactive selection of a long-term Delta strategy to serve the state's residents and the Delta ecosystem better than the deteriorating status quo?

A peripheral canal, combined with mitigation for loss of Delta farmlands, could protect water quality for agricultural and urban export users as well as in-Delta urban users. It also would be compatible with more variable salinity conditions for fish. Because a canal would not be able to provide all Delta farmers with a substitute source of fresh water, it might be most practical—whether or not it is legally necessary—to develop a complementary program to provide transitional assistance to affected Delta farms.[8] As long as everyone agreed, it might be possible to negotiate the necessary changes in Delta water quality standards. But with holdouts, the problem might be difficult to resolve without legal action.

The state must lead in resolving these conflicts by taking a forward-looking view of changing water quality conditions and needs. The Board has the legal authority and the tools to take the lead on this effort, although it lacks the resources, political support, and mandate to do so. The Board recently resolved to develop a multiyear strategic work plan on Delta issues. This is an opportunity to consider future regulatory frameworks that can work best for the ecosystem and the state's economy.[9] A more proactive Board can push the regulatory discussion with federal, state, and local officials to find realistic ways to live with the changing conditions and uses of Delta waters. Delta salinity is the first of many such issues that California will face as the climate warms. For example, in-stream temperature standards on many rivers and streams, including many within the Delta watershed, are also regulated under the Clean Water and Porter-Cologne Acts, and it may become increasingly challenging to meet these standards with longer warm seasons and warmer inflows into reservoirs.

ANTICIPATING LEVEE FAILURES

As highlighted in Chapter 3, the physical forces acting on the Delta suggest an increasing likelihood of levee failures in the coming decades, and for many islands the costs of repair may well exceed the value of economic activity and infrastructure assets that the levees protect. Similarly, the modeling results in Chapter 5 suggest that only the western islands might be important for maintaining Delta salinity standards. These findings suggest that it will not be in the interests of Delta landowners or the state to repair all levees after failures, and that it may also be in the state's interest to develop a strategy for purchasing and preflooding some islands to reduce salinity intrusion from extensive levee failures.

Clearly, additional economic analysis and hydrodynamic modeling work is needed to map out a long-term levee strategy of this type. Important legal issues also need to be considered regarding the potential hydraulic effects of island flooding on landowners on neighboring islands. These effects can include greater wave action and increased underseepage, requiring reinforcement of the neighboring levees to avoid higher flood risk. We estimate that these mitigation costs can be substantial, ranging from several million to more than 10 million dollars per island, depending on the size of the flooded island and the length of levees affected on neighboring islands.[10]

There is no explicit statutory requirement to mitigate changes to neighboring levees if a levee breaks. In this case, neighboring landowners would need to resort to tort law and prove that the levee owner was negligent or deliberately caused the levee failure. Even if fault were found, it might be difficult to receive payment from the local reclamation districts responsible for nonproject levees, because under the terms of Proposition 218 (a constitutional amendment passed in 1996), the districts would not have funds unless island landowners voted to assess themselves. Flooded landowners are unlikely to have the will or the capacity to do so, particularly for islands that are not to be repaired.

The situation is likely to be quite different if the state is directly involved, and the issues differ for nonproject and project levees. For nonproject levees, the state might purchase islands either as part of a long-term mitigation strategy for Delta landowners or with the intent to pre-flood the islands. In the first case, the state would likely be more exposed financially than private landowners, even if it did not deliberately cause the islands to flood. Preflooding islands might make the state liable for consequences to neighboring islands. In short, the state needs to develop a policy regarding neighboring island levees if it gets into the business of buying Delta lands.[11]

The state does not currently have the option of not repairing project levees after a failure without the agreement of the U.S. Army Corps of Engineers—an action that would likely require congressional approval. Thus, any forward-looking policy regarding project levees—some of which protect islands highly at risk—must anticipate these issues and involve federal consultations well in advance.

INCLUDING UPSTREAM DIVERTERS IN A DELTA SOLUTION

Most reductions in net Delta outflow are due to upstream diversions and consumptive use of surface water and groundwater (see Chapter 4). In an average year, water users upstream of the Delta on the Sacramento and San Joaquin rivers and their tributaries divert roughly twice the amount of Delta water as export users. Although the Board has broad authority to include upstream diverters in meeting environmental water quality needs in the Delta, efforts to do so have been limited.

In 1986, the Racanelli Decision (*United States v. State Water Resources Control Board*, 227 Cal Rptr. 161, at 195–1986) clarified that all water rights holders, irrespective of seniority, could be required to participate in meeting water quality standards. The decision made it clear that the Board has the authority to set water quality standards for beneficial uses including, specifically, protection of fish and wildlife. The Environmental Impact Report for the 1995 WQCP examined several alternatives for placing some responsibility for Delta water quality standards on upstream diverters (State Water Resources Control Board 1999). The two alternatives that allocated responsibility by order of priority resulted in relatively little participation by upstream diverters, because most have rights senior to the export projects. A third alternative projected a much broader sharing of responsibilities, by relying on proportional cutbacks in upstream diversions on a watershed basis, irrespective of seniority. In the end, the CVP and SWP assumed responsibility for the water quality standards, but deals were made to seek contributions from senior agricultural users on the San Joaquin and Sacramento rivers, in exchange for financial compensation.[12] In separate actions, the Board has required upstream diverters to modify flows to meet the needs of fish populations in the tributaries themselves.[13]

Upstream diversions could increase greatly in the future (Whitney 2008). Potential avenues include perfection of the so-called "state filings"—water rights applications filed by the Department of Finance to reserve priority rights for other users when the CVP and the SWP were built. In addition, upstream water users in the "areas of origin" can receive higher priority for new water rights applications. Presently, over four million acre-feet of water rights applications are pending in the Delta watersheds; most (if not all) would rely on area of origin claims for seniority over the projects.[14] By comparison, Delta exports, at their maximum, averaged roughly six million acre-feet per year.

The potential for new upstream diversions, even if limited to a portion of the applications on file, raises questions about the long-term reliability of current planning efforts for Delta exports. One alternative to offset greater upstream diversions would be to move from a priority-based approach toward a watershed-based approach, with proportional cutbacks, for regulating water quality. Such an approach might be most consistent with the Public Trust Doctrine. Another would be to increase the use of market-based tools, building on existing arrangements to get senior upstream diverters to release flows in exchange for compensation. As noted in Chapter 7, there is considerable potential for increasing outflows through a combination of higher minimum outflow regulations and market-based mechanisms.

Any reduction in upstream surface water diversions in the Sacramento and San Joaquin basins might provide little additional flows into the Delta unless it was accompanied by actions to limit expansions of groundwater use. Many upstream users, when faced with reduced access to surface water, can merely shift water demands to groundwater. Given the hydraulic connection between surface water and groundwater in these basins, additional consumptive use of groundwater in the Sacramento and San Joaquin basins eventually leads to reductions in surface flows or groundwater mining. This physics of the problem is essentially unrepresented in California groundwater law (Sax 2002).

PROTECTING ENDANGERED SPECIES IN THE FACE OF UNCERTAINTY

An alarming aspect of the current Delta crisis is the decline of native fish populations, several of which are listed under the federal and state Endangered Species Acts. Judge Wanger's recent decision to curtail pumping to protect delta smelt was a remedial action under federal endangered species law and will significantly reduce exports below those allowed under the WQCP (see Chapter 2). The current efforts to develop a Bay Delta Conservation Plan (BDCP) reflect water export users' goals to move to a more flexible regulatory regime for species protection. The BDCP is being designed to serve jointly as a Natural Communities Conservation Plan (NCCP)(under a state law that complements the state Endangered Species Act) and a Habitat Conservation Plan (HCP) under Section 9 of the federal Endangered Species Act. Within a NCCP/HCP framework, the export users would move from being regulated on a species-by-species basis, with incidental "take" permits for harm done to species, to a regime in which the overall conservation plan for a group of species guides regulatory intervention. With a plan that is sufficiently protective of the stated conservation goals, which must include species recovery under the terms of the NCCP, the export users hope to have assurances that they will not face the type of cutbacks that have occurred under the Wanger ruling.

An NCCP may provide the most promising process for dealing with aquatic species management issues in the Delta; it lays out clear guidelines for conservation goals, supported by scientific review, and it is the only statute that explicitly considers adaptive management as part of the conservation process.[15] Developing such a plan for the Delta will be challenging, given the number of players and the complexity of aquatic habitat and water operations issues. To date, other NCCPs have focused on terrestrial habitat protection, and the "project" at stake is where to allow land development—a relatively straightforward issue, with fewer moving pieces.[16]

Even with an approved plan, BDCP participants will likely continue to face some legal and regulatory uncertainty, judging by the NCCP experience in Southern California.[17] In the Delta, there is also a persistent risk that some species will not do well, even if the plan's conservation actions are well designed and carried out in earnest. The results of our expert survey show that the scientific community has serious doubts about the viability of delta smelt under any water management alternative, even under the best cases (Figure 6.1). With climate change, the chances of viability decline significantly for this and other key Delta species. In addition to the many existing stressors, water temperature increases will make it harder for some species to find a suitable window of time to spawn and thrive.

Neither the state nor federal Endangered Species Acts account for the possibility of losing a species due to climate change. Like the Clean Water Act, these laws were passed in the 1970s, well before climate warming was in the spotlight, and they assume that harm to species in a project area is caused by direct human action. As a result, some important questions have not yet been tested: Can a well-planned NCCP/HCP protect against loss of a species from an external event such as climate change? Would incorporating climate change effects in the plan's adaptive management program—to foster the best conditions for the species—be adequate to provide coverage?

Even if a species declined solely because of climate change and the Endangered Species Act did not apply, it may be difficult to argue that CVP and SWP operations are not exacerbating or hastening the risk of extinction. Given the extent of physical manipulation of water in the Delta, proving that the projects play no role is difficult. Thus, Endangered Species Act enforcement could still shut down or significantly reduce exports, as long as there is a reasonable chance that diversions are contributing to the problem. Issues are likely to arise for fish other than delta smelt, as evidenced by Judge Wanger's April 2008 ruling concerning winter- and spring-run Chinook salmon (see Chapter 2). Those involved in the planning process need to take this risk into account when evaluating various alternatives and their costs.

Under current law, the only recourse to a direct conflict between species and economic losses would be a congressional exemption to the Endangered Species Act for the Delta, or a favorable ruling from the "God Squad"—an interagency cabinet-level group that can exempt projects from the act if the economic costs of compliance are too high. These are high-stakes events; to date, exemptions have been granted in only a handful of cases.[18] Here, as with the Clean Water Act, the Delta's issues are acute but not unique: Numerous terrestrial and aquatic species are at risk of extinction from climate-related changes in habitat, accentuating the trade-offs between species protection and economic development (Davis and Shaw 2001; Malcolm et al. 2006).

GOVERNANCE SAFEGUARDS FOR A PERIPHERAL CANAL

Among the export management alternatives considered in this book, two would involve constructing a peripheral canal. Because this alternative would be a major departure from the present system of diversion, it would require new governance mechanisms. As noted earlier, the peripheral canal is highly controversial. In June 1982, the last time a peripheral canal was seriously considered, it was rejected by an overwhelming majority of Northern California voters (Figure 2.4). The two main concerns are still being voiced by some today: the potential for a "water grab" by Southern California and the effects of a canal on the Delta ecosystem.[19]

Although the San Francisco Bay Area now depends on the Delta as much as urban Southern California does, Sacramento Valley residents are sensitive to how much water can be exported from their watershed without causing local economic harm. And although there are potential environmental benefits from changing the intake points for water exports, environmentalists want to ensure that enough water is made available for habitat needs in the Delta if export water is diverted upstream.

One way to satisfy these apprehensions would be to provide physical safeguards, for example, by building a very small canal or pipeline. However, given the variability of rainfall and the scale economies of canal sizing, this solution would limit the economic benefits of improving the conveyance of water exports. For several reasons, a very small canal also risks limiting the environmental benefits: (1) It would not allow diversions to vary over the course of the tidal cycle, increasing risks of entraining downstream organisms; (2) it would make it more difficult to allow salinity to vary within the Delta and Suisun Bay; (3) it would limit flexible, adaptive operations that might reduce entrainment of fish at export intakes and (4) it would encourage prolonged, substantial pumping from the southern Delta, which is harmful to fish (see Chapter 5). A better solution is to build a canal large enough to benefit from water management opportunities and to provide solid safeguards through the governance and regulatory system.

Providing safeguards to Sacramento Valley residents is largely a political issue, although considerations of "safe yield" to the region's groundwater basins could also play a role in setting export limits. The problem could be readily dealt with by setting long-term average limits to Delta exports—for example, at the average of the past 10 or 20 years. This period would

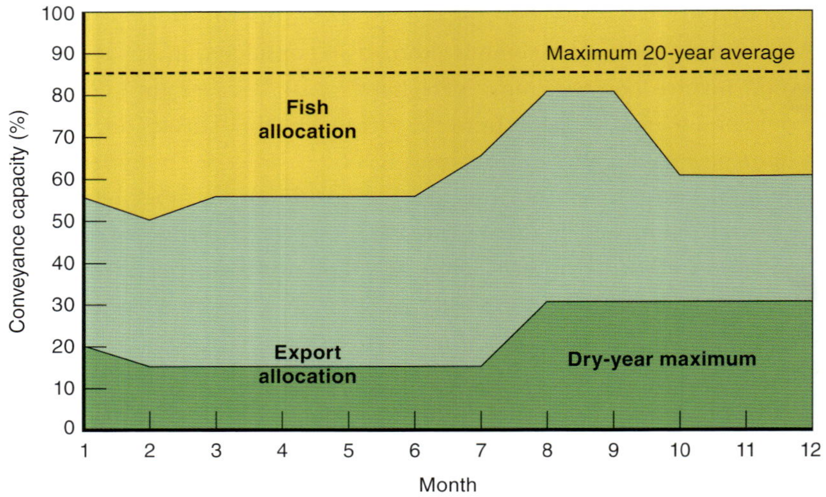

FIGURE 8.2 Allocation of peripheral canal capacity in a system with safeguards. Additional Delta outflow requirements also would exist (not shown here). Annual and monthly percentages shown here are for illustration only. In dry years, export users could convey only the volume represented in the dark green area. The fish allocation (yellow area) would be available for various Delta environmental uses or for lease to export users, when the water was not needed for fish. Over a 20-year period, exports would not be allowed to exceed a maximum 20-year average (or other appropriate threshold), to protect interests in the Sacramento Valley watershed.

need to exceed the common decadal periods of wet and dry years. Such limits could be instituted by regulations, ownership of long-term capacity, or surcharge fees dedicated to environmental restoration or water development in Northern California.

Providing safeguards for the ecosystem requires substantial scientific input. In addition to guaranteed minimum inflows into the Delta for ecosystem needs, the ideal system would provide the ecosystem with variable flows across seasons and years, depending on conditions of the fish and other factors. To allow for this flexibility, a formal Delta Environment Authority might control a sizable amount of conveyance capacity, which could be allocated to Delta inflows, to lower San Joaquin River flows, or leased to export users, depending on ecosystem needs. For some period of time, the minimum inflow requirement could include adequate flows to maintain salinity standards for in-Delta diverters, until this latter goal became unattainable because of sea-level rise or island flooding. Export users, too, would have a lower bound of water availability from the canal, which would vary seasonally and by water-year type. Hydrodynamic modeling and analysis by biologists could help establish the size and pattern of these allocations.

Figure 8.2 provides a simple illustration of such a system. In practice, one might launch the system with a higher allocation for environmental flows, and ramp up export allocations as fish populations improved. A side benefit of this flexible arrangement is that leasing of the fish allocation on some occasions could create a stream of income for ecosystem investments.

Some parties could still worry that the system could be undone through the political process—for example, by a change in the laws governing the canal or the public institutions that manage it. To provide legal safeguards, two alternative approaches have recently been proposed. The first, suggested in the 2007–2008 legislative session, is to provide a constitutional protection of export limits.[20] An alternative proposal is to consider a type of public-private partnership for managing the canal, with a private party (for example, an environmental water trust) to manage the flexible allocation for the ecosystem.[21] With a private partner, the governance rules for canal operation would be subject to private contracts law. If the agreement specified appropriate compensation for abrogation of the contract terms, this could make the system less vulnerable to modification by administrative or legislative fiat. Effective legal safeguards for environmental flows have occurred elsewhere in California, such as with Mono Lake, increased Trinity River flows, and the protection of Wild and Scenic Rivers (Hundley 2001).

GOVERNANCE AND DECISION-MAKING FOR A NEW DELTA

The CALFED experience of the 1990s and early 2000s shows that stakeholder processes cannot be relied on to make major strategic decisions for the Delta. Too many divergent interests were involved and essentially any interest could block any major decision, effectively limiting actions to modest and largely ineffective modifications of the status quo. Today, prospects for stakeholder decision-making are further dimmed because diminishing state and federal funds reduce external incentives for agreement. The urgency and magnitude of the Delta's problems require more capable frameworks for making strategic decisions. The transition to a new Delta will require a fundamental reorganization of the Delta's governance and regulatory framework. The legislature and governor, in consultation with local governments, stakeholders, and the federal government can best undertake this task. The state attorney general's office could play a pivotal role by assessing available legal and institutional options.

Historically, Californians have made major strategic decisions regarding water: for flood control in the early twentieth century and for development of major projects in the middle of the twentieth century (see Chapter 2). In both cases, decisions were preceded by long periods of controversy. But persistent crises and the mounting need for strategic change ultimately prevailed in effecting change. These decisions reconfigured existing local governments and state and federal agencies to implement fundamentally new directions in water management. Absent the effort to make comparable decisions today, Delta management will remain in the realm of tinkering with a deteriorating status quo until court decisions or physical catastrophe intervene.

However, affirming a strategic decision alone is insufficient. Broad policy directions for water management established by the state must include real institutional, financial, and technical capability and authority. Establishing such directions, in a state with many competing problems and few available funds, will require leadership and financial involvement from the beneficiaries of its implementation.

CONCLUSION

In sum, although opportunities exist to improve the economic and environmental outcomes in the Delta, innovative solutions will face significant legal and regulatory hurdles. The first issue is the inflexibility of the Clean Water Act. Sea-level rise, climate change, the needs of the Delta ecosystem, and water quality and reliability concerns for water export users are all pushing in the direction of more variable Delta salinity, which could preclude some present agricultural uses. The State Water Resources Control Board will need to work with federal officials to see how California can establish necessary regulatory changes to Delta water quality standards, while complying with federal law.

To build a peripheral canal, which could provide numerous water quality and reliability benefits, it will be necessary to overcome concerns about the unreliability of current legal protections for the environment and upstream users. Many of the safeguards these parties seek could be provided through a governance structure that ensures a flexible allocation of water for the ecosystem and limits long-term export volumes from upstream basins. This would allow the size of the canal to be based on optimal water management opportunities for both human uses and the Delta ecosystem, rather than on fears that too much water might be diverted.

Current planning processes will need to incorporate the continued risk of water export cuts, even if a canal is built. To seek greater regulatory certainty, export users are currently pursuing a more comprehensive approach to habitat protection and species recovery in the Delta within an NCCP/HCP framework. However, the risks to species are high, and there are unanswered questions about how such a plan would protect water export projects if species continue to decline, as long as diversions can be linked to the problem. These risks will increase with climate change and the associated rise in water temperatures. In addition, water export projects face cuts from increased diversions in upstream watersheds, which are senior in priority under area of origin laws. Regulatory and market approaches will have to be pursued to lessen this risk.

The state also will need to engage in active planning to anticipate the changes in Delta landscapes with the increased risk of island flooding. Some islands may not be worth repairing based on their economic values, and a policy of preflooding some islands may be warranted to limit the risks of catastrophic failure. If the state develops a policy to acquire Delta lands—either to ease transitions for Delta farmers or to facilitate preflooding—it must also consider the potential costs to neighboring island levees that could be affected by island flooding. Forward-looking consultations with federal agencies are also required to develop new policies regarding the project levees that form part of federally authorized flood control projects. In addition, land development decisions will need to come under greater regulatory review to consider both flood protection and the environmental consequences of land development.

The transition to a new Delta will require a fundamental reorganization of the Delta's governance, finance, and regulatory framework. The legislature and governor, in consultation with local governments, stakeholders, and the federal government can best undertake this task. No Delta solution exists that will be good for all parties. This was a delusion of the CALFED era, borne of a now-depleted state and federal cash flow. Reaching a political agreement in the face of trade-offs will be difficult and will likely require some compensations and mitigations. Such mitigations will require either greater external (state and federal) funding or increased payments from beneficiaries. In any event, beneficiaries will almost certainly need to pay most of the costs of fixing the Delta.

9

DECISION ANALYSIS FOR DELTA EXPORTS

> The secret of getting ahead is getting started. The secret of getting started is breaking your complex, overwhelming tasks into small manageable tasks, and then starting on the first one.
>
> MARK TWAIN (1835–1910)

The Delta poses a variety of highly complex problems with a myriad of uncertainties. These troublesome characteristics are common to many other problems, ranging from public policy issues such as national defense and school system planning to personal career and retirement planning. To address all aspects of such problems simultaneously is beyond human abilities and comprehension. To solve complex problems, it is first necessary to organize them into smaller components that can be understood and solved sequentially, to provide insights into how to solve other pieces and to indicate promising overall strategies.

In this chapter, we organize recent scientific and technical findings and assessments summarized earlier and elsewhere to evaluate each export management strategy with respect to the environmental and water supply objectives for the Delta, measured in terms of native fish population viability and statewide economic costs of water supply. In evaluating these strategic decisions, it is important to recognize that not everything is known (or can be known) and that uncertain future events will require responses. The analysis presented here aims to incorporate uncertainties explicitly, by considering ranges of values for costs and other outcomes, as well as responses to some major potential problems and opportunities. In this way, this analysis provides a basis for evaluating alternatives despite uncertainties. Society can rarely afford to make decisions without uncertainty. In the Delta, postponing a strategic decision because of uncertainty amounts to making a decision to continue the deteriorating status quo.

DECISION ANALYSIS APPLIED TO THE DELTA EXPORT ALTERNATIVES

To make this economic and ecosystem assessment, we employ formal decision analysis, incorporating the costs and opportunities of things going wrong and things going well for each export strategy. Decision analysis formalizes the evaluation and comparison of decision options with uncertainty and multiple objectives. Although some aspects of this approach date to the early 1800s, decision analysis as practiced today was largely an outgrowth of the development of operations research during World War II. Today, decision analysis methods are taught widely in business and engineering schools, commonly taught in economics and public policy, and increasingly applied to environmental and natural resource problems (Cohon 1978; de Neufville 1990; ReVelle et al. 1997; Hobbs et al. 2004; Maguire 2004; Lund 2009). These methods are widely employed for problem-solving in business and engineering.

A detailed discussion of the approach, method, and assumptions of this formal analysis appears in Lund et al. (2008b). From a statewide economic point of view, major choices and outcomes for Delta export management strategies can be depicted as in Figure 9.1.

The box at the left-hand side represents the initial decision: to use a peripheral canal intake, to employ "dual" export facilities, to end water exports, or to continue through-Delta pumping. In the figure, the simplest choice to represent is to end Delta exports, which results in a direct and relatively certain cost (as discussed later), and is summarized or valued in a box on the right-hand side of this option.

The decision to build a peripheral canal is more complex, because it is uncertain how a canal will affect the major fish species of concern for the Delta. In its simplest form, this uncertainty can be represented as two possible outcomes: Either the fish recover or they do not recover (as judged by biological, political, or legal standards). This uncertainty cannot be resolved until the canal has been built; it is represented in the decision tree by a circle with two chance events: fish recovery or failure of the fish to recover. In the happy event of fish recovery, operation of the peripheral canal is

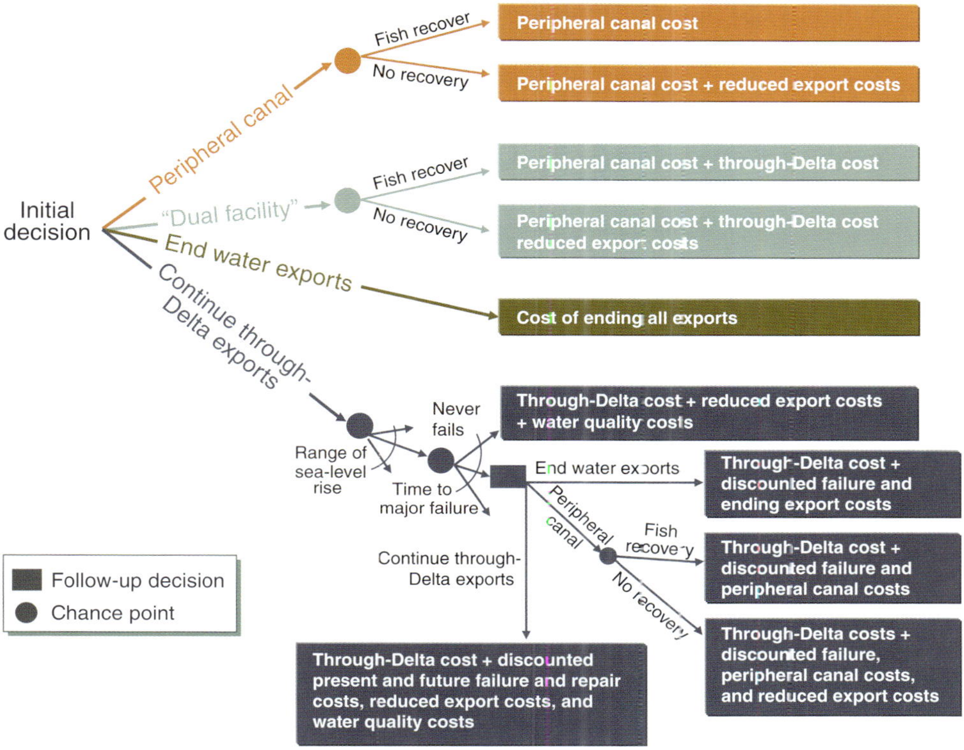

FIGURE 9.1 Decision tree for strategic long-term Delta exports from a statewide economic perspective.

relatively unfettered. In this case, the cost of the canal with fish recovery is represented in the box on the right-hand side as the cost of the peripheral canal. However, if fish populations do not recover, exports from a peripheral canal would likely be subject to legal and political pressure for substantial cutbacks relative to the levels of the recent past (pre-Wanger decision), which would incur substantial economic costs. As discussed in Chapters 6 and 8, there are biological and legal reasons to expect legal and political pressure to reduce exports from a peripheral canal. The costs of reduced peripheral canal exports are included in the boxed costs for this combination of choice and chance event. The expected cost for a peripheral canal would then be the average of the costs for these two chance events, with each event cost weighted by our assessment of the probability of each outcome.

A dual conveyance alternative combines a peripheral canal with continued through-Delta pumping. This alternative, as we represent it, combines the costs and probabilities of a peripheral canal with additional costs for continued use of through-Delta pumping. Other more complex representations of dual conveyance alternatives could be employed; but at this stage, the range of dual conveyance proposals is still poorly defined, making a simple representation most appropriate. For example, the current representation of dual conveyance cost does not include water quality costs or damages from extensive levee failures.

The most complex choice is to continue exclusively with through-Delta pumping. Here, the major chance event is an extensive failure of Delta islands, which is a function of the rate of sea-level rise and other physical drivers of change discussed in Chapter 3. For each uncertain rate of sea-level rise, there is a different probability of extensive Delta island failures. This is represented by a second circle, representing different years before an extensive levee failure occurs. If the Delta never experiences extensive levee failure, then the cost is the sum of costs for through-Delta facilities, the costs of any reduced exports from continued through-Delta pumping, and additional water quality costs for urban and agricultural areas receiving Delta water, which is more saline than water that would be drawn into a canal further upstream on the Sacramento River (see Chapter 7).

When there is an extensive failure of Delta levees, a follow-up decision (known as a "recourse" choice) must be made—to repair and continue through-Delta pumping, to end Delta exports, or to construct a peripheral canal. These choices have a structure similar to those discussed earlier. The total cost for each recourse choice includes not only the costs of the recourse choice, but also the by-then "sunk" costs of through-Delta facilities and the damage costs of extensive Delta island failure.

INFORMATION NEEDED FOR DECISION ANALYSIS

Following this structure of the decision problem, from a statewide economic perspective, a series of 16 questions must be answered to complete the analysis. The 16 questions appear in Table 9.1, along with our suggested range of answers. These questions take into account future sea-level rise (#1), the likelihood of extensive Delta levee failure and how it varies with sea-level rise (#2 and 3), the likelihood of being able to maintain viable fish populations with different intake strategies (#4 through 7), the

TABLE 9.1

Decision Analysis Questions and Answers Recommended by the Authors

Consideration	*Low Value*	*High Value*
Sea-level rise (ft)		
1. How much will sea level rise by 2050?	0.5	1.5
Probability of extensive Delta levee failure by 2050 (annual failure probability in parentheses)(%)		
2. With the minimum sea-level rise?	34 (1)	88 (5)
3. With the maximum sea-level rise?	57 (2)	95 (7)
Population viability in 2050 for delta smelt (Chinook salmon in parentheses)(%)		
4. What is the likelihood of viable fish populations with continued through-Delta pumping?	5 (10)	30 (30)
5. What is the likelihood of viable fish populations with no Delta exports?	30 (40)	60 (80)
6. What is the likelihood of viable fish populations with a peripheral canal?	10 (20)	40 (50)
7. What is the likelihood of viable fish populations with dual conveyance?	10 (20)	40 (50)
8. By what proportion would exports be reduced for fish protection with continued through-Delta pumping? (%)	25	40
9. If the fish do not recover, by what proportion would peripheral canal water exports be reduced? (%)	25	40
Economic and financial costs ($ billion)		
10. What is the construction cost of a peripheral canal?	4.75	9.75
11. What is the additional drinking and agricultural water quality cost of Delta water?	0.3/yr	1.0/yr
12. What is the annualized cost of ending Delta exports?	1.6/yr	2.5/yr
13. What is the annualized cost to maintain through-Delta pumping?	0.15/yr	0.4/yr
14. What is the cost to water users of a sudden extensive failure of Delta levees?	7.8	15.7
15. What is the average cost to repair an extensive Delta levee failure for water supply?	0.2	2.5
16. What exponent relates export reduction to economic cot?	2	3

SOURCE: Lund et al. 2008b.

reductions in exports likely for each strategy if the fish are not recovering (#8 and 9), and the economic costs of implementing each strategy and the costs of failures or reduced or ended water exports (#10 through 16).

To better represent the uncertainties and many other complexities of the problem, we offer both high and low values for each parameter. Fish population viability estimates are undertaken for two key Delta species. For delta smelt, viability is defined as achieving sufficient recovery to avoid Endangered Species Act restrictions on water exports. For fall-run Chinook salmon, it is defined as maintaining adequate populations to support commercial and recreational fisheries. The values chosen for the analysis are based on the available technical and scientific evidence, the authors' own experienced professional judgment, as detailed in Lund et al. (2008b), and by other technical studies which support that work (Hanak and Lund 2008; Suddeth et al. 2008; Fleenor et al. 2008; Moyle and Bennett 2008; Tanaka et al. 2008; Chen et al. 2008; Medellin-Azuara et al. 2008b). A summary of our reasoning for selecting the values for each answer appears in the appendix to this book, with more detail in Lund et al. (2008b).

The use of ranges of costs and probabilities implicitly captures additional uncertainties that are not explicitly included here. This formulation of the problem, although relatively simple, is sufficiently rigorous and understandable to provide insights into desirable choices for the Delta intake decision. (Occasionally, a more rigorous formulation becomes less understandable and obscures any resulting insights, sometimes called "rigor mortis.")

TABLE 9.2 *Annual Costs and Likelihood of Fish Population Viability for Delta Export Alternatives, 2050*

Delta Export Alternative	*Average Cost ($ billion/year)*	*Likelihood of Viable Populations (%)*	
		Delta Smelt Population	Fall-run Chinook Salmon Fishery
Continuing through-Delta	0.55–1.86	5–30	10–30
Peripheral canal	0.25–0.85	10–40	20–50
Dual conveyance	0.25–1.25	10–40	20–50
No exports	1.50–2.50	30–60	40–80

SOURCE: Lund et al. 2008b.

COMPARING WATER EXPORT ALTERNATIVES

Spreadsheet calculations were used to determine the statewide economic costs and probabilities of fish population viabilities for each of the four alternatives using the decision analysis framework presented in Figure 9.1 and the ranges of estimates provided in Table 9.1. Table 9.2 summarizes the results of these calculations. Figure 9.2 presents these same results graphically.

For some alternatives, the range of likely economic performance is quite broad, reflecting uncertainties about cost. In particular, for continued through-Delta pumping, the expected costs range from as low as $550 million to nearly $1.9 billion per year. A key uncertainty for this option is how soon the system will be damaged by a large-scale levee failure. For fish, the ranges reflect the considerable uncertainties about species performance, depending in part on how carefully the ecosystem components are managed, as well as on influences from external sources (i.e., for salmon, the ocean and upper watershed).

Despite these uncertainties, some clear comparisons emerge. In terms of statewide economic cost, the most likely ordering of alternatives is peripheral canal (best), followed by dual conveyance, continued through-Delta

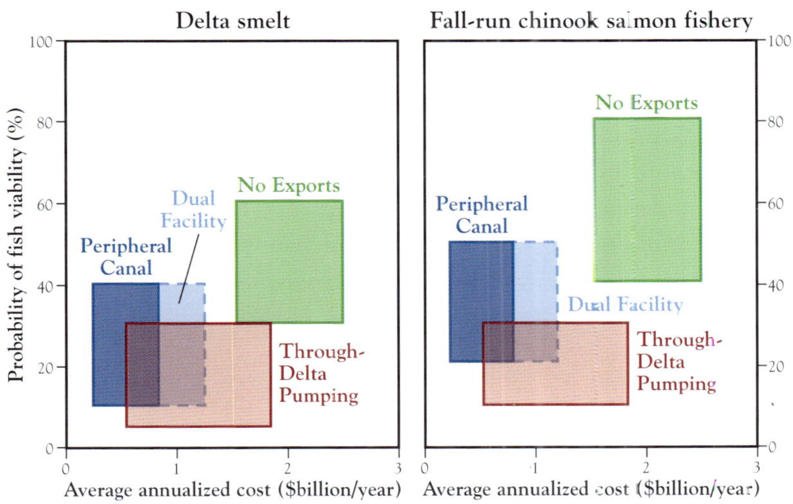

FIGURE 9.2 Range of costs and fish population viability for Delta export alternatives in 2050. Source: Lund et al. 2008b.

pumping, and last, ending all exports. Even with relatively high construction costs (on the order of $10 billion) and 40 percent pumping cutbacks to support endangered fish species, the costs of a peripheral canal are below $900 million per year.

Dual conveyance is potentially more costly, because of additional infrastructure costs to maintain the viability of through-Delta pumping. Costs also could be somewhat higher in this alternative because of the increased water quality costs for urban and agricultural users of the portion of water pumped through the Delta and the costs of repairing extensive Delta levee failures.

Several key drivers lead to higher costs for continued through-Delta pumping. First, by mid-century, the water quality costs of taking water from the Delta are on the order of $300 million to $1 billion per year. Second, this alternative requires significant investments, initially to fortify key levees and perhaps also to improve Delta channels and ultimately build a peripheral canal when the levee system fails.[1] Third, a catastrophic failure of key levees would cause a large one-time cost of $8 billion to $16 billion.

The no-exports alternative, in contrast, involves considerable costs outside the Delta itself, as water users develop alternative, higher-cost sources and reduce agricultural and urban use, particularly for agriculture in the southern Central Valley.

The most likely ordering is quite different for fish viability. The no-exports alternative is best, followed by peripheral canal and dual conveyance systems (tied), and continued through-Delta pumping last. There is a broad consensus among estuarine experts that ending exports is likely to be best for a range of desirable fish species (see Chapter 6). Benefits include ending the harmful entrainment and unnatural flow patterns generated by southern Delta pumps, as well as providing more water for aquatic habitat. A peripheral canal also provides the first of these benefits if it is designed and operated to minimize new entrainment problems at the upstream intake.

Although in principle a dual conveyance alternative offers some additional flexibility for water management, we do not believe that it will have appreciably different outcomes for either delta smelt or salmon compared to a pure peripheral canal alternative. Finally, continued through-Delta pumping is least beneficial for fish, given the problems of entrainment and disruption created by southern Delta pumps. Through-Delta pumping also prevents the more flexible management of environmental water flows to increase aquatic habitat area and variability. Eventually, through-Delta pumping, even at reduced levels, will lead either to the elimination of exports entirely or to the construction of a peripheral canal (the least expensive recourse after extensive failure). In this context, it is worth noting that "opportunistic" through-Delta pumping—taking water only when the Delta was sufficiently fresh—does not appear to be a viable long-run alternative. With sea-level rise and island flooding, usable Delta water will become less frequent, less reliable, and of poorer quality. Opportunistic pumping also will be limited by environmental constraints, because the freshest flows (generally in winter and spring) tend to occur at times when pumping cutbacks may be needed to protect desirable fish.[2]

How do the alternatives compare when environmental and economic performance are considered together, as coequal objectives?

- The peripheral canal and dual conveyance alternatives are very likely to perform better than continued through-Delta pumping on both objectives. We calculate that a peripheral canal has a two-thirds chance of outperforming through-Delta pumping on both economic and fish objectives; for dual conveyance, the chance is 60 percent (see Lund et al. 2008b). In contrast, through-Delta pumping has only a 5 percent chance of outperforming the two canal-based alternatives on both objectives.
- We find little technical reason to prefer dual conveyance over a peripheral canal. The two alternatives are likely to perform equally from a fish perspective, and dual conveyance is likely to be more costly. Nevertheless, for an interim period, it may be valuable to maintain through-Delta pumping as part of a dual system, to maintain water quality for Delta farmers and provide transition flexibility for exports and environmental operations.
- A clear trade-off exists between a peripheral canal and dual conveyance and the alternative of ending exports. Peripheral canal and dual conveyance costs are lower, whereas ending exports is better for fish. Selecting between these alternatives requires a value judgment.

Are the policy options limited to a choice between (1) a peripheral canal and dual conveyance approach with moderate probabilities of viable fish populations and lower economic costs or (2) ending exports with very high costs and somewhat higher probabilities of maintaining fish populations? No. Hybrid approaches exist where some savings from a peripheral canal and dual conveyance approach are invested in habitat and restoration activities to benefit fish. Alternatively, reducing the volume of exports from a peripheral canal or dual conveyance alternative also might improve fish populations. The former approach may provide the most useful compromise, because it provides a key to proactive investments in habitat improvement, which are likely to be needed under any export management alternative (see Chapter 6).

How stable and robust are these conclusions? We have tested them by varying the range of answers and making some changes in the calculation methods. In general, these conclusions seem to be robust enough to address the uncertainties and complexities not included explicitly here (Lund et al. 2008b). Even substantial reduction of water exports (below pre-Wanger decision levels) results in a peripheral canal being preferable to reduced exports. But others are welcome to provide their own estimates (hopefully with technical justifications) to test these conclusions. The spreadsheet provided with Lund et al. (2008b) is designed to allow users to modify the answers to the 16 questions and see how the results change.

IMPLEMENTATION ISSUES

We find that there is a substantial scientific and technical basis for making a policy decision on the strategy for water exports from the Delta. However, a host of major implementation issues remain for guiding the creation of a new Delta, including Delta island policies; governance, regulatory and finance institutions; operations; and ecosystem management (see sidebar Design and Operations Options for a Peripheral Canal).

Although the physical forces driving the Delta and the economic analysis presented in Chapter 3 and Suddeth et al. (2008) indicate that it will be uneconomical and ultimately impossible to maintain all Delta levees, Californians have only begun to discuss potential policies for Delta islands, such as which islands should be repaired, how failed islands should be managed (i.e., converting some islands to aquatic or terrestrial habitat),

DESIGN AND OPERATIONS OPTIONS FOR A PERIPHERAL CANAL

INFRASTRUCTURE DESIGN

- Upstream intake locations
- Additional intake locations
- Outlet locations
- Total flow capacity
- Fish screening
- Sedimentation basin
- Booster pumping
- Right-of-way
- Channel elevations and lining
- Stream channel crossing
- Associated operational water storage
- Associated recreational facilities

MAJOR ADJUSTMENTS AND MITIGATIONS

- Delta farmers
- Contra Costa Water District
- North Bay Aqueduct
- Delta towns
- Recreation
- Environment

OPERATION POLICIES

- Operating strategy
- Constrained delivery policies
- Monitoring

DELTA LAND AND WATER MANAGEMENT

- Aquatic and terrestrial habitat
- Flood management
- Levees
- Agriculture
- Recreation

GOVERNANCE, REGULATION, AND FINANCE

- Ownership
- Governance authority
- Regulatory oversight
- Finance and repayment
- Terrestrial and aquatic habitat management

SOURCE: Lund 2008.

and so on. As noted in Chapter 8, these decisions also raise important legal and regulatory questions regarding levee policy. A systematic and comparative examination of Delta island and land use policy is needed from a realistic long-term perspective, with accompanying policy discussions and decisions.

Likewise, governance, finance, and regulatory regimes for water quality, instream flows, and endangered species also will need to be redefined in the transition to the new Delta (see Chapter 8). Regulations will need to be compatible with federal requirements. Governance and finance are not abstract activities; they must be designed to accomplish and implement a general strategy for the Delta.

This book largely avoids detailed discussions of water operations because of the short time frame and insufficient capability to include such detailed analysis. Systematic study of operational issues will be needed over an indefinite period, even many years after any new Delta policies have been implemented, given changing problems and understanding of the Delta. This will require substantially new and different types of hydrodynamic, operations, and planning analysis. Many months of analysis will be needed to inform discussions of preliminary operating policies for policy and planning purposes and negotiations.

The design and implementation of ecosystem management activities is perhaps the most important and difficult area where additional implementation work is required. For decades, California has neglected the synthetic scientific thinking and difficult policy discussions required to develop a sustainable vision of the kind of ecosystem that can and should be maintained in the Delta. A quantitative analysis capability that assesses how water operations mesh with ecological objectives is needed to better inform discussions involving trade-offs among water export and environmental objectives.

Beyond these general areas, specific, detailed implementation issues must be resolved over the course of policy, planning, design, construction, and operations for the new Delta. The sidebar highlights the types of decisions required for a peripheral canal alternative; similar lists could be developed for the other three alternatives examined in this book. These issues all require an ability to make and implement policy decisions. New Delta governance and finance arrangements must be capable of making and implementing such detailed decisions, within a broader policy framework established at the legislative and state executive level.

Most of these decisions would be aided considerably with additional scientific and technical information, which still needs to be developed or assembled from previous studies. The Delta's transition will bring Californians into unfamiliar territory, where intuition and an understanding based on how things have operated in the past will become less reliable predictors. Only scientific and technical analysis can help guide the way through this new landscape.

THE TIMING OF DELTA DECISIONS AND CONSEQUENCES

Another aspect of Delta decision-making that we have not considered in detail is timing—for instance, how one might phase in a new export management regime. To provide some insight, Figure 9.3 presents a conceptual view of how export alternatives may perform over time and the choices Californians will face. Water exports are currently being curtailed from historical high levels as a result of court rulings regarding endangered species (and the ongoing drought). Additional species listings are likely to cause further export reductions in the near term. The accumulating effects of land subsidence, sea-level rise, worsening floods, and earthquakes will make continuation of through-Delta pumping less reliable and more costly over time, but will leave peripheral canal exports relatively unaffected. However, it will take some time before a peripheral canal can be constructed. How well a dual conveyance alternative ultimately performs

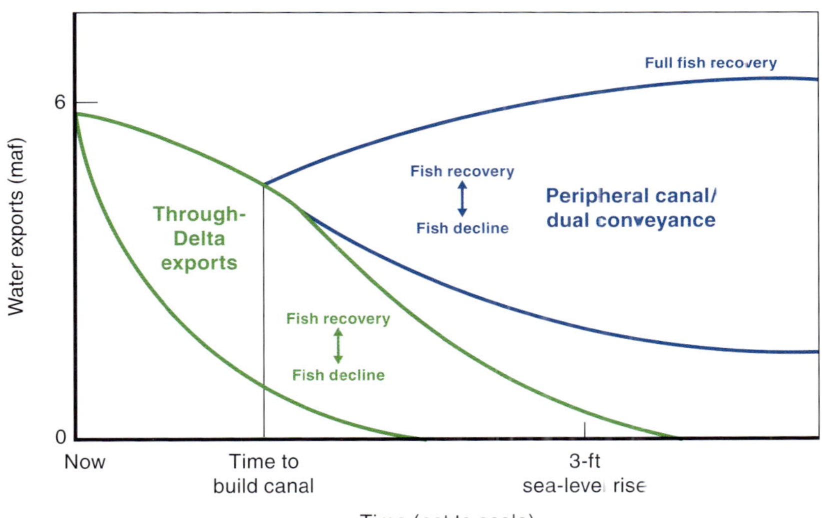

FIGURE 9.3 Delta export transitions over time.

will depend on the size of the canal component. If the canal is sufficiently large, it can take an increasing share of exports as through-Delta pumping becomes less viable. As Figure 9.3 highlights, the ability of each alternative to support fish populations also significantly affects its ability to support exports.

CONCLUSION

We developed and applied a formal decision analysis framework to examine four long-term strategies concerning water exports from the Delta. The options examined include (1) pumping water through the Delta (the current policy), (2) taking water exports around the Delta through a peripheral canal, (3) combining through-Delta pumping and a peripheral canal (dual conveyance), or (4) ending Delta water exports altogether. The analysis considers two main criteria for performance, consistent with two primary objectives for the Delta: ecosystem revitalization and water supply. We measure ecosystem revitalization by the yardstick of viability of two desirable Delta fish populations and water supply by the yardstick of economic costs of water supply and water quality. We focus on outcomes for the middle of this century. This is a sufficiently long horizon to incorporate the effects of natural forces acting on the Delta, such as sea-level rise, and yet close enough in time to be relevant for today's decisions about major infrastructure investments.

Our results suggest that continued use of through-Delta pumping is risky and unlikely to be the best alternative from either a statewide economic perspective and an environmental perspective. After an extensive set of levee failures in the Delta, it will be less costly to replace through-Delta pumping with a peripheral canal than to rebuild the through-Delta system. Building a canal sooner, before an extensive levee failure, is less costly to the economy. A proactive policy may avoid the high costs of an abrupt interruption of water supplies and might provide significant water quality savings and public health benefits. A peripheral canal also is likely to be better for a variety of desirable fish species. A dual conveyance alternative has similar prospects for Delta fish, at potentially higher costs.

Ending Delta exports entirely is the most favorable strategy for maintaining the viability of desirable fish populations. However, it comes with the greatest statewide economic costs and would deprive environmental management in the Delta of a potential revenue source. The hundreds of millions of dollars of lower average annual costs from the peripheral canal and dual conveyance strategies provide a statewide resource for environmental investments in the Delta. Redirecting some of this economic gain to habitat acquisition and other environmental activities might improve the viability of desirable fish species. Reducing exports at times might have a similar function, at an economic cost. To succeed in meeting economic and environmental goals, Californians will need a more coherent program of operational management for the new export facilities, strongly coordinated with habitat management, than has been present in export management programs to date.

10

CHARTING THE FUTURE FOR A CHANGING DELTA

> Once a landscape has been established, its origins are repressed from memory. It takes on the appearance of an "object" which has been there, outside us, from the start.
>
> KARATANI KOJIN (1993), *Origins of Japanese Literature*

To be successful, natural resources management must be able to adapt to changing conditions. This book has looked at the long-term management of California's Sacramento–San Joaquin Delta, which faces inevitable changes in landscape, economy, and ecology, driven by sea-level rise, climate change, earthquakes, land subsidence, and biological invasions. Management objectives for this region have also been changing over time, as a consequence of long-term shifts in the societal values placed on the Delta's ecosystem and the species that depend on it to thrive. Now we summarize our conclusions regarding the Delta's changing landscape, the potential for improving the Delta's ecosystem, the alternatives for managing water exports from the Delta, and the regulatory and governance challenges that lie ahead. Each of these issues must be addressed as part of a proactive management strategy for the new Delta.

THE CHANGING DELTA LANDSCAPE

Fundamental changes are inevitable for the Delta. "Restoring the Delta" is an unrealistic and perhaps meaningless notion given the historical changes that have occurred in the Delta and the immutable forces that will operate on it for decades to come. The Delta of the future will have fewer islands and more open water.

Increasing Risks of Island Failures

Sea-level rise, earthquakes, continued land subsidence, and higher winter flood flows will increase the frequency of Delta island failures and the costs of preventing and recovering from failures. Under today's risk conditions, most of the Delta's subsided islands have at least a 90 percent chance of failing some time in the next 50 years. These drivers of change, including an escalating threat of earthquakes and sea-level rise of approximately one foot by 2050 and 3 feet by 2100, will significantly increase this likelihood of failure.

High Costs of Prevention and Recovery

Maintaining all Delta islands is not cost-effective. Reducing the frequency of island flooding in the Delta would cost many billions of dollars. From a water supply perspective, only the western Delta islands might be essential for keeping salinity away from export pumps in the southern Delta (before significant sea-level rise brings salinity farther into the Delta in any event). Continued investment in many islands in the northern, southern, and eastern Delta can be supported by the economic value of on-island activities and infrastructure such as roads and rail lines. But for nearly 20 subsided Delta islands, there is no compelling economic basis for state investments in levee upgrades or in repairing and restoring the islands after failure.

Inevitable Shift to More Open Water

Given the magnitude of projected change during this century, it is unreasonable to assume that the current levee network will be maintained indefinitely at increasing costs and diminishing benefit. These costs, coupled with increasing risk factors, ensure that the Delta landscape of the future will be significantly different from the Delta of the past. Within the next 50 years, the Delta very likely will contain large areas of open water left after islands have flooded. Water quality will be substantially degraded for water export purposes, and perhaps improved for desirable fish species.

A Need for Readiness

California is unprepared for the changes that will occur in the Delta. Current institutions, regulations, infrastructure, and expectations for the Delta are built around maintaining the Delta in an unsustainable and deteriorating condition. It is time to prepare for a very different Delta, with a different ecosystem and different water supply and land use capabilities. With timely, purposeful action, there is some choice in what the Delta will become. California will need institutions with sufficient authority, finance, focus, and leadership to accomplish this transition successfully.

FISH AND THE DELTA ECOSYSTEM

Promising opportunities lie ahead for improving conditions for desirable fish and wildlife in the Delta. For fish, there is bound to be improvement in aquatic habitat as more is created by island flooding. Changes in water operations and habitat management can improve conditions not only for fish but also for other wildlife, especially waterfowl.

Benefits of Island Flooding

In recent years the Delta ecosystem has shifted to a less-suitable state for desirable fish species. Large-scale flooding of Delta islands is likely to create habitat that is no worse and probably better for most desirable fish. Besides expanding the extent and volume of aquatic habitat in the Delta, large-scale flooding will greatly alter water movement through the Delta. The suitability of the new open-water habitats for desirable species will depend in part on the responses of harmful invasive species, including overbite clam and Brazilian waterweed, to the changed system. A proactive experimental approach is required to guide the evolution of habitat in these flooded areas—they will be larger and deeper than the currently flooded islands (e.g., Franks Tract), which are poor examples of the future landscape.

Need for More Diverse Aquatic Habitat

More diverse habitat is fundamental to improving conditions for desirable fish, and greater variability in Delta water flow and quality is part of this strategy. Changes in Delta water management that allow for greater spatial and temporal variability in water flows and quality (salinity, turbidity, etc.) could improve conditions for native fish species while making conditions less favorable for invasive species. In addition to increasing the variability in water conditions, actions to benefit desirable species should include increasing the extent of floodplain and tidal marsh habitat within the Delta. Major opportunities to create such diverse habitat conditions exist in the northern Delta (Cache Slough region and Yolo Bypass), Suisun Marsh, and other areas.

Role of Water Exports

Water export alternatives matter for fish. The current system of through-Delta pumping is the least desirable alternative from an environmental perspective. In addition to killing some fish at the pumps, the present system alters flow patterns within the Delta, moving desirable species to undesirable habitats. A peripheral canal could reduce these problems, while allowing Delta waters to be managed for greater variability. Dual conveyance—combining a peripheral canal with continued through-Delta pumping—may offer opportunities to avoid killing fish under some circumstances. But overall, dual conveyance is not likely to be better for fish than a peripheral canal operated on its own. Eliminating exports entirely is the most promising alternative for key species, including delta smelt, longfin smelt, and the four runs of Chinook salmon. However, careful management of water exports with a peripheral canal or dual conveyance—with substantial complementary ecosystem investments—can significantly improve the compatibility of continuing exports and rebuilding viable populations of desirable species.

Essential Ecosystem Investments

Rebuilding large, self-sustaining populations of desirable Delta fish species will require large and carefully designed ecosystem investments. No matter which water export option is adopted in the future, large investments are needed for habitat acquisition, restoration, and improvement. Likewise, increases in scientific knowledge are needed to effectively manage desirable species (native fish species and others that do well under similar conditions). Delaying these investments will increase their costs, reduce the likelihood of fish population recovery, and raise chances of greater water export reductions.

Unavoidable Species Risks

The prospects for some Delta species are not good, even if society does everything possible to help them, as fast as possible. For example, delta smelt's very survival is threatened by rising water temperatures (from climate change) on top of all the other factors. The Delta is also likely to continue to be a poor environment for juvenile Chinook salmon under most likely scenarios, increasing the difficulty of saving the listed spring and winter runs of Chinook salmon and of sustaining commercial salmon fisheries. The potential for losing some species over the next 50 years poses great environmental, legal, and regulatory challenges.

LONG-TERM WATER EXPORT ALTERNATIVES

For water exports, time favors a peripheral canal and is unfavorable to alternatives that rely on through-Delta pumping. Although ending exports entirely would be best for fish, a peripheral canal is an unavoidable component of a long-term solution that serves both economic and ecosystem objectives. A dual conveyance system is, at best, an interim solution (Table 10.1).

Increasing Costs of Sea-level Rise

Sea-level rise will make through-Delta pumping increasingly unattractive. Even if the existing levee network could be maintained through unprecedented investments, worsening Delta water quality from sea-level rise will steadily reduce the economic value of water exports from within the Delta. The current costs of Delta salinity are already significant for southern Central Valley agriculture and urban drinking water treatment. More saline Delta exports will reduce the viability of agriculture in this region and increase costs of and health risks from drinking water from the Delta. Alternatively, higher salinity will impose a direct water supply cost by requiring higher outflows to repel seawater from the pumps. With 3 feet of

TABLE 10.1 *Summary Comparison of Water Export Alternatives*

Alternative	*Role*
Continued through-Delta exports	Increasingly unstable and costly solution
Dual conveyance	Interim solution for transition to peripheral canal
Peripheral canal	Potential to provide both cost-effective water supply and improved fish viability
No exports	Best for fish but most costly to the economy; ultimate outcome without a peripheral canal

sea-level rise—a possibility by late in this century—through-Delta pumping may no longer provide a major source of fresh water without large increases in Delta outflows, even if the western islands can be kept intact. Even "opportunistic pumping" export alternatives which involve taking water from the Delta only when flows are freshest, will become less frequent, less reliable, and of poorer quality. Opportunistic pumping also will be limited by environmental constraints, because the freshest flows (generally in winter and spring) tend to occur at times when pumping cutbacks may be needed to protect desirable fish.

Two Ultimate Export Alternatives

Over the long term, the choice of water export strategies narrows from four to two: building a peripheral canal or ending Delta exports. Given its unreliability, increasing costs, and environmental risks, transporting water from Northern California through the Delta to other parts of the state is not a viable long-term option. So the choice comes down to diverting all exports around the Delta or ending exports and making do with other supplies in regions currently relying on exports.

Although ending exports would provide significant tangible benefits for desirable fish, this strategy would be particularly burdensome for the state's economy. It would also likely increase the difficulty of raising the financial resources necessary for environmental investments in the Delta and prolong the period when water exporters would attempt to divert water from the Delta, to the detriment of desirable fish species.

A peripheral canal would provide significant benefits to the regions relying on exports. In addition to water supply and quality benefits for urban users, there are potentially important benefits to agriculture and the environment. Reducing the salinity of water exported for agriculture might greatly extend the economic life of agriculture in the southern Central Valley, and should eventually provide some improvement in San Joaquin River salinity. If properly managed, a canal could significantly improve conditions for desirable Delta fish relative to the present export system. The weaker environmental performance of a peripheral canal compared with ending exports might be offset by channeling some of the economic surplus generated by the canal to enhance ecosystem investments.

Compared with a peripheral canal, dual conveyance does not offer much added environmental promise, but it can help maintain water quality for farmers in the southern Delta under modest levels of sea-level rise and will likely be a necessary interim solution. In the very near term, some investments will be needed to maintain the deteriorating through-Delta system as the transition is made to ending exports or building a peripheral canal. The transition away from through-Delta pumping will occur over time, whether planned for or not. A more expeditious, planned transition would be less susceptible to the rapid, costly changes accompanying earthquakes, floods, and levee failures. A well-planned expeditious transition is also likely to benefit desirable fish species.

GOVERNANCE, REGULATION, AND FINANCE

A successful Delta solution will require governance, regulatory, and financial mechanisms and institutions that allow firm decisions to be made in a timely way. This institutional framework must include contributions, involvement, and responsibility of water export users and also should include upstream diverters (who remove almost twice the amount of Delta outflows as export users) under a broader statewide authority directed towards a firmly established strategic policy.

Institutional Challenges of New Conveyance

By making it possible to divert water around the Delta, a peripheral canal creates opportunities for environmental and economic benefit, but it also raises new institutional challenges. One issue is whether to provide safeguards for the environment and other water users by limiting the size of the canal or devising an iron-clad governance system. A second is financing: Even if export users agree to pay for the canal—as they have indicated they would—funds must be raised for ecosystem investments and to mitigate harm to Delta farms whose water quality conditions could deteriorate more quickly with a canal. A third is whether the regulatory system can be adapted to the new and changing conditions. The governance and regulation of a peripheral canal would need to fall within a larger governance and regulatory strategy for Delta water and land use management.

Safeguards Through Governance

Governance mechanisms can be devised to provide appropriate safeguards for a peripheral canal. Northern California's concerns that a canal would export too much water from the region can be met by setting long-term maximum export levels, enforced by regulations and law, surcharge fees, or capacity ownership. Environmental safeguards for adequate instream flows can be provided by allocating a share of capacity to the environment, which can be used as needed or leased to fund restoration efforts. For legal and political reasons, water exports through a peripheral canal probably should begin at reduced levels, and increase as fish populations improve. With such safeguards, it should be possible to build a canal large enough to take advantage of tidal flows and California's variable hydrology for both environmental and economic purposes.

A User-based Financing System

User-based financing mechanisms are available to cover a wide range of water system needs. The precedent of having export users pay for their own infrastructure costs is well established with the State Water Project; this model should be extended to any new conveyance facility. Because export users will benefit directly from more reliable and higher-quality water, and because exports will continue to cause some environmental problems, it is also appropriate for users of a peripheral canal to pay an eco-surcharge on export volumes. Some portion of the significant water quality cost savings of a peripheral canal could be allocated to environmental programs. Upstream diverters, who currently account for nearly two-thirds of all water consumption from the Delta watershed, should also be expected to financially support ecosystem programs. Ecosystem finance would also benefit from the ability to lease shares of conveyance capacity from environmental owners. Finally, some public funds may be appropriate to supplement these sources and to help cover mitigation costs for in-Delta users, although such funds are unlikely to be plentiful given the long-term financial problems facing the state and federal governments.

Unprecedented Regulatory Challenges

The regulatory framework as it stands today is not prepared to oversee the Delta of the future. Neither the Clean Water Act (1972) nor the Endangered Species Act (1973) acknowledges the effects of climate change, a key driver of future Delta conditions. Under the terms of the Clean Water Act, it may be difficult to take proactive steps to protect Delta exports from encroaching salinity and to increase habitat variability by building a peripheral canal, because this may hasten the natural decline of water quality for some Delta farmers. The Endangered Species Act could make it difficult to develop a reliable long-term habitat conservation plan for the Delta, given the risks of extinction for some species under a changing climate and the difficulty of disentangling the role of water exports from species decline.

Similarly, land use regulation in the Delta needs more attention to prevent inappropriate urbanization that endangers public safety and precludes environmental protection. Fortunately, successful models exist elsewhere in California, such as the Coastal Commission and the Tahoe Regional Planning Agency. These agencies use permitting authority to balance economic and environmental goals in environmentally sensitive areas. To develop a planned transition to return some Delta islands to aquatic habitat, the state also must resolve important legal issues regarding liability for levee failures.

The Frontline Role of Science

The implementation of an effective Delta strategy will require an unusual amount of scientific and technical involvement. The new Delta is *terra incognita* in terms of its biological response, water quality, and water management operations. Initial water and habitat management and operations policies are unlikely to be fully satisfactory, and will need to be adapted over time. To garner relevant field data and expertise, a commitment to continuous improvement in water and habitat management is essential. Technical and scientific efforts for managing the new Delta will require more independence from political processes than in the past.

NAVIGATING CHANGE

The Delta of the future will be very different, and the costs of inaction are high. California needs to prepare for this changed future to direct it more favorably. This means charting a new strategic direction for Delta water management, planning environmental investments from the vantage point of a changing Delta, and developing strategies to manage levee failures. Preparing for the future through governance and regulatory reforms is also essential.

The ongoing and increasingly rapid changes in the Delta pose a long-term challenge to California as a whole, as well as to all parties involved in this perennial source of conflict. All parties seeking to achieve both environmental sustainability and water supply reliability have an interest in making a peripheral canal part of a long-term solution for the Delta. This strategy must be embedded in a broader set of actions to improve aquatic environments in the Delta and the greater watersheds of the Sacramento and San Joaquin rivers. To be viable, a long-term solution must include governance, regulatory, and financial arrangements to ensure that various goals are well served, including water supply, environmental management, and the state's local interests in the Delta. It is unlikely that local and regional stakeholders can negotiate such arrangements on their own in a timely way, given the complexity of the problem and its innumerable stakeholders. Pursuit of a grand consensus solution for the Delta's many issues is likely only to continue the deteriorating status quo. Leadership from the governor and legislature will be needed to create conditions and direction for reasonable governance of the new Delta, with cooperation from local governments and federal agencies that regulate and manage water and land use.

APPENDIX

ESTIMATION OF PROBABILITIES, COSTS, AND REDUCTIONS FOR DELTA OUTCOMES AND STRATEGIES

> Errors using inadequate data are much less than those using no data at all.
>
> CHARLES BABBAGE (1792–1871)

This appendix summarizes the scientific and technical rationales and discussion used to provide low and high estimates for answering the 16 questions in Table 9.1. The interested reader can find more detailed explanations, as well as discussions of major assumptions, technical limitations, and sensitivity analysis in Lund et al. (2008b).

SEA-LEVEL RISE

1. *How much will sea-level rise by 2050?*

Warming temperatures are increasing sea-level rise globally in two ways. First, a warming atmosphere causes the ocean to warm, which expands the volume required to contain a given mass of water (a process known as "thermal expansion"). Second, warmer temperatures melt continental ice sheets and glaciers, adding water to the ocean that was previously stored in these reservoirs of ice. In California, records suggest a past rate of sea-level rise of 10 to 20 centimeters (4 to 8 inches) per century, which is similar to the global estimate. The rate of global sea-level rise has accelerated in recent years with potential for much greater sea-level rise over this century.

Drawing on research by Cayan et al. (2008a), Bindoff et al. (2007), Meehl et al. (2007), and Rahmstorf (2007), the CALFED Independent Science

Board has recommended that the Delta Vision effort use midrange values for sea-level rise of 8 to 16 inches by 2050 and 28 to 39 inches by 2100 for planning purposes (Mount 2007). We use figures similar to these as our lower (0.5 feet) and higher (1.5 feet) bounds for sea-level rise for 2050 (although it is the failure probability ranges associated with sea-level rise, rather than direct sea-level rise, that drive our calculations).

PROBABILITY OF EXTENSIVE DELTA LEVEE FAILURE

For this decision analysis, we consider extensive levee failures of the western islands from earthquakes or floods, since others are less important for water supply (see Chapter 5 and Fleenor et al. 2008). Recent reports highlight the importance of the poor foundations of western island levees and their general proximity to several earthquake faults (Mount and Twiss 2005). The draft DRMS report (URS and Jack R. Benjamin and Associates 2007b) finds seismic risk for western islands is exceptionally high, with annual failure probabilities as high as 1-in-20 (5 percent per year) for each island. Failure rates are likely to increase with sea-level rise (Suddeth et al. 2008).

Preemptive flooding of large, highly subsided, and vulnerable islands in the interior of the Delta could reduce their potential to contribute to failure of Delta water supplies in a major earthquake. We have assumed this is not done.

2. *With the minimum sea-level rise, what is the probability of extensive Delta levee failure by 2050?*

To establish a low range value, we assumed that all five western islands would receive substantial upgrades to make them resist earthquakes or be more easily repaired following an earthquake at a cost of $3.6 to $5.2 billion for such upgrades (URS and Jack R. Benjamin and Associates 2007b). We assumed this construction would reduce both seismic risk and risk due to low rates of sea-level rise or modest increases in inflows, creating a level of protection for all five islands (not each individual island) approximately equal to an urban standard of 1 percent annual probability of failure, or 34 percent cumulative probability of failure by 2050.

For the high range of failure with minimal sea-level rise, inflows, and earthquakes, we used something less than the median level of combined annual flood and seismic failure probability outlined in Suddeth et al. (2008): 5 percent annual probability of failure, for a cumulative probability of failure of 88 percent by 2050. These values are well below the estimated current annual risk of failure of the western islands suggested by the DRMS study, and are an upper bound of an optimistic view of failure likelihood.

3. *With the maximum sea-level rise, what is the probability of extensive Delta levee failure by 2050?*

Here we used the range of failure probabilities outlined by Suddeth et al. (2008) for the entire Delta. That is, the best possible failure probability under rapid rates of sea-level rise, high seismic risk and increasing inflows is approximated by the island with least current combined risk of failure from flooding and earthquakes at 2 percent per year, for a cumulative probability of failure of 57 percent by 2050. For the high end of risk, we used the current estimated combined flood and earthquake annual failure probability of the western Delta islands of approximately 7 percent per year, or a cumulative probability of failure of 95 percent by 2050 (Jack R. Benjamin and Associates 2007b). These estimates are based on *current* probabilities of failure. Even this high end number may not capture the increase in risk with time (Suddeth et al. 2008).

FISH POPULATION VIABILITY IN 2050

What is the probability of viable fish populations with:

4. *continued through-Delta pumping?*
5. *no Delta exports?*
6. *a peripheral canal?*
7. *dual conveyances?*

There is no broadly accepted method for estimating the probability that Delta fish populations will remain viable under different water management scenarios in the future, and many factors affect population viability.

Nevertheless, a considerable body of scientific knowledge now exists about the role water operations play in the declines of key pelagic species in the Delta—the native delta smelt and longfin smelt, and the non-native striped bass (e.g., Bennett 2005; Feyrer et al. 2007). Water operations are also recognized to affect the health of the salmon runs that migrate through the Delta on their way to and from the ocean (Moyle 2002; Kimmerer 2008). Here, we rely on expert judgment to assess the ranges of viability.

For this exercise, we estimated the probability of fish population viability for two key Delta species that have fairly different requirements. For the delta smelt, viability is defined as achieving sufficient recovery to avoid Endangered Species Act restrictions on water exports. For fall-run Chinook salmon, it is defined as maintaining adequate populations to support commercial and recreational fisheries. While delta smelt can actually be driven to extinction by water management, fall-run Chinook salmon (the principal run) are likely to persist in small numbers under most scenarios, but the valuable recreational and commercial fishery for them may disappear.

In addition to our team's own biologists, we sought input on these questions from a sizable group of experts on the Delta ecosystem. The results of this survey are provided in Bennett et al. (2008), and they are quite consistent with our own judgments (Moyle and Bennett 2008).

Our own rankings of fish population viability under different export alternatives largely coincide with those from the expert survey (see Figure 6.1). However, an important difference concerns the likely outcomes for salmon with a peripheral canal, which we believe could be significantly better than with continued through-Delta exports. We recognize that a peripheral canal will need to be designed and operated to minimize entrainment and other potentially negative effects. We assume that solving these problems will be part of the price of a peripheral canal solution. A major reason why a peripheral canal could improve conditions for Sacramento River salmon is that it would allow Sacramento River water to flow straight through the Delta, without diversions into the central and southern Delta. For San Joaquin salmon, ending through-Delta exports would allow salmon to avoid entrainment and high predation rates from the water export pumps. Likewise, with through-Delta pumping, juvenile salmon will most likely continue to suffer considerable mortality from being carried to less desirable places within the Delta (e.g., the central Delta) and by increasing predation by resident fish species, as their movement through the Delta is slowed or as they are brought into the vicinity of the pumping plants.

Another difference between our rankings and those from the expert survey concerns the relative performance of the peripheral canal-only and dual conveyance alternatives. Although the dual conveyance alternative is potentially a more flexible water management tool from a fish perspective, we do not believe that it will make a significant difference in the long run for either the pelagic organisms, such as delta smelt, or for anadromous fish, such as the Chinook salmon. Our main reason is that dual conveyance would have to be operated at both ends on very short time scales (hours or days), based on knowledge of the presence or absence of desirable fish. While this might be possible, the history of water project operations suggests that such a rapid and flexible operation to favor fish is highly unlikely. Thus, the dual conveyance strategy may suffer from the disadvantages of both the peripheral canal-only and through-Delta pumping strategies.

8. *By what proportion would exports be reduced for fish protection with continued through-Delta pumping?*

With continued through-Delta pumping, we estimate that water exports are likely to be reduced in the long term by between 25 and 40 percent. This is slightly higher than the anticipated reductions resulting from the recent federal court (Wanger) decision, which limits reverse flows on Old and Middle rivers to protect delta smelt.[1] The Department of Water Resources has estimated that this decision would result in export cuts of 22 to 30 percent in an average water year for State Water Project customers (Department of Water Resources 2007b). The U.S. Fish and Wildlife Service's new biological opinion for delta smelt, released on December 15, 2008, may lead to increased restrictions (U.S. Fish and Wildlife Service 2008). Our estimate of higher potential cutbacks, of up to 40 percent on average, reflects the possibility of additional restrictions. In particular, longfin smelt was listed as threatened under the California Endangered Species Act in March 2009. In April 2008, Judge Wanger ruled that the biological opinion for the coordinated operations of the Central Valley Project, State Water Project, and the Operational Criteria and Plan (OCAP) is invalid for several listed anadromous fish species (winter and spring-run Chinook salmon and Steelhead), likely resulting in additional remedies and restrictions on water exports (see Chapter 2).

9. *If the fish do not recover, by what proportion would peripheral canal water exports be reduced?*

If desirable fish species do not recover despite this change in intakes, water exports are likely to continue at reduced levels, despite the presence of a peripheral canal. We estimate that these reductions would be similar to the 25 to 40 percent cutbacks currently imposed on through-Delta pumping.

This estimate is based on several factors. First, delta smelt, longfin smelt, and Chinook salmon all depend on freshwater inflow to the estuary for rearing and (for the smelt) spawning during the late winter and spring (as reflected in water quality regulations to maintain the salinity gradient, X2, at the mouth of the Delta from February through June). Second, the period of highest vulnerability (February through June) is also when outflows are likely to be highest, with the most water available for upstream diversions. Third, more freshwater flows and reduced pumping also may be needed in the fall to provide habitat for delta smelt in upper Suisun Bay and the Delta. Thus, if the fish do not recover, windows of opportunity for use of a peripheral canal are likely to be confined to months when water is least available.

It is unclear how the likely range of export reductions would differ between the peripheral canal-only and dual conveyance strategies. The greater operational flexibility might reduce the size of export reductions, but the cutbacks also could be very similar to those of a peripheral canal, because the periods when more through-Delta pumping could occur are also times when effects on fish are likely to be greatest. The calculations assume that the cutbacks for dual conveyance would take the average value of cutbacks for through-Delta and peripheral canal-only alternatives.

ECONOMIC AND FINANCIAL COSTS

Our cost estimates all assume similar levels of investment in environmental improvements to those of the recent past. Greatly increasing environmental investments could improve fish population viability, albeit at greater economic cost. Ecosystem investments might vary in content, cost, and effectiveness for different alternatives and implementations.

10. *What is the construction cost of a peripheral canal?*

Our range of costs for a peripheral canal is $4.75 to $9.75 billion, based on the following inputs. For the canal itself, we use estimates developed in 2006 by the Metropolitan Water District of Southern California. A canal on the eastern side of the Delta was estimated to cost on the order of $4 billion, with small variations depending on whether the canal was routed entirely in the primary zone of the Delta or partly in the secondary zone. A western alignment, drawing water from the Sacramento Deep Water Ship Channel and including some portion with deep tunneling, ran on the order of $7.5 billion. We inflate these cost estimates by 20 to 30 percent to allow for cost overruns.

The estimates include fish screens, investments for seismic safety (for a 200-year seismic event), and resilience to at least 3 feet of sea-level rise. (The western alignment would be resilient to higher levels of sea-level rise, given the greater ease of moving intakes further upstream). Estimates for both alignments also include the costs of building an additional forebay, consistent with the maintenance of dual conveyance, at roughly $0.1 billion. This cost could be eliminated for a peripheral canal only.

11. *What is the additional drinking and agricultural water quality cost from using Delta water?*

We anticipate that additional drinking water treatment costs from Delta water would fall in the range of $0.1 to $0.7 billion per year, possibly continuing as low as they are today (if combined with increased Delta outflows) or much higher with additional salinity intrusion and tighter drinking water standards (Chen et al. 2008). The long term costs of Delta salinity for southern Central Valley cropping and animal production (relative to using Sacramento River water) appears to be an additional $0.2 to $0.3 billion per year, even without additional salinity intrusion from sea-level rise and island failures (Medellin-Azuara et al. 2008b) Combining drinking water and agricultural water quality costs, the range of total water quality costs from through-Delta exports as opposed to upstream diversions is roughly $0.3 to $1.0 billion per year.

12. *What is the annualized cost of ending Delta exports?*

For a low cost estimate for ending exports, we use the CALVIN estimate of $1.5 billion per year. For a high cost of ending exports, we use $2.5 billion per year. This includes the cost estimate of ending exports without allowing additional transfers from the Tulare Basin and adds an additional $300 million per year, for reasons described below. These estimates result in an average cost of roughly $250 to $420 per acre-foot of foregone exports (Tanaka et al. 2008).

The $1.5 billion per year cost is probably an optimistic (low) estimate because it comes from an optimization model, which assumes that water users have perfect hydrologic foresight and no difficulties reaching agreements to transfer and exchange water. However, CALVIN may miss some opportunities for cost savings if the model is unaware of less expensive options that water users may be able to develop. The most likely errors that would decrease costs of ending exports would be lower costs for water conservation. Ending additional water transfers from the Tulare Basin to Southern California cities raises the costs of ending exports about $700 million per year to $2.2 billion per year (about $370 per acre-foot). Transfer restrictions in agricultural regions are a likely consequence of a no-export scenario, as agricultural counties seek to avoid further local economic losses. The sum of $300 million per year is added to this estimate to reflect the likely optimism of the CALVIN model.

13. *What is the annualized cost of maintaining continued through-Delta pumping?*

To continue to operate the Delta as a water supply conveyance will require upgrading the western islands to resist earthquakes. The DRMS Phase 2 report estimates the cost of upgrading eight western islands to resist an earthquake with a recurrence interval of 300 years (an annual probability of 0.3 percent) is on the order of $8.1 billion. Here we use a slightly lower range of $3 to $5 billion, assuming only five western islands to be critical. The annualized cost of this investment, at a 5 percent discount rate, would range from $150 million to $250 million. These costs seem necessary for prolonged use of Delta intakes for water exports.[2]

Another potentially important cost for maintaining through-Delta pumping is the additional carriage water needed to keep the Delta fresh enough to support exports. With sea-level rise this amount will increase. For one foot of sea-level rise, the average increase may be on the order of 400,000 to 500,000 acre-feet per year, with considerable variation across years (Fleenor et al. 2008). If exports are valued at $100 to $200 per acre-foot, this cost ranges up to $50 million to $100 million per year.[3] We omit these costs in our estimates of maintaining through-Delta pumping, because we have already accounted for the costs of declining water quality with sea-level rise (see Question #11).[4] However, for small increases in required Delta outflow, it might well be less expensive to reduce exports than increase treatment costs.

Summing the costs of annual Delta levee maintenance and seismic upgrades of the five western islands gives a range of annualized costs by 2050 of approximately $165 to $280 million per year to maintain the Delta as a viable source of fresh water for pumping.[5]

As an alternative to the approach to fortifying the western islands, some proposals and cost estimates have been made to fortify Middle River levees for seismic resistance at a cost of roughly $500 million. This seems to be an interim measure, perhaps also useful for longer-term dual conveyance operations. Other estimates include from $1 to $10 billion in capital cost to modify channels within the Delta to create and maintain a fortified through-Delta canal (Department of Water Resources 2008). If viewed as an alternative to the reinforcement of western islands examined earlier, this through-Delta reinforcement would result in a much wider range of annualized costs, from a low of $67 million to a high of $528 million, without maintenance costs.

Overall, the lowest costs seem to be about $150 million per year for inexpensive Delta improvements with increased carriage water, plus $30 million per year for maintenance costs. The highest costs would be about $400 million per year for maintaining the western islands and increasing carriage water, with $80 million per year for Delta improvements and other expenses.

14. *What is the cost to water users of a sudden extensive failure of Delta levees?*

An "extensive" failure of the Delta would involve failure of a sufficient number of islands critical to water exports during a single event, ending exports to water users outside of the Delta for up to two years. The DRMS Phase 1 report calculated the economic costs and impacts of an event of this type. DRMS examined three different conditions during this event—wet spring, average summer, dry fall—and calculated the economic costs of emergency response and repair, infrastructure repair, lost use of structures and services, and agricultural and recreation losses within the Delta. The calculated estimate—roughly 80 percent of the total statewide economic costs—would come from disruption to water users, with the balance resulting from the inability to use some infrastructure. The range of economic costs to water users is $7.8 to $15.7 billion, depending upon conditions during failure and the length of time it takes to restore water deliveries. The additional costs to repair levees and islands to restore water exports are addressed separately in the following question.

15. *What is the average cost to repair an extensive Delta levee failure for water supply?*

For this analysis, we estimated a range of values for an event which would lead to flooding of five western islands critical to water supply. For the low-end cost, we assume the levees have already been upgraded to seismically repairable levels and that there is only a single breach in each island. Based on the DRMS estimates of damage and repair costs for a "mean higher high water" flood of these five islands, the total low-end cost would be approximately $184 million.

High-end costs are more difficult to estimate and depend principally upon the length of levees to be rebuilt. If a major earthquake struck the western Delta before seismic upgrades of levees had been completed, multiple levee failures would likely occur on each island, requiring extensive repairs taking several years. (Then again, repairs may not be possible given the domino effect of levee failure once a few large islands are flooded). The DRMS study evaluated the repair costs of a variety of failure scenarios, although none precisely match the simultaneous loss of all five western

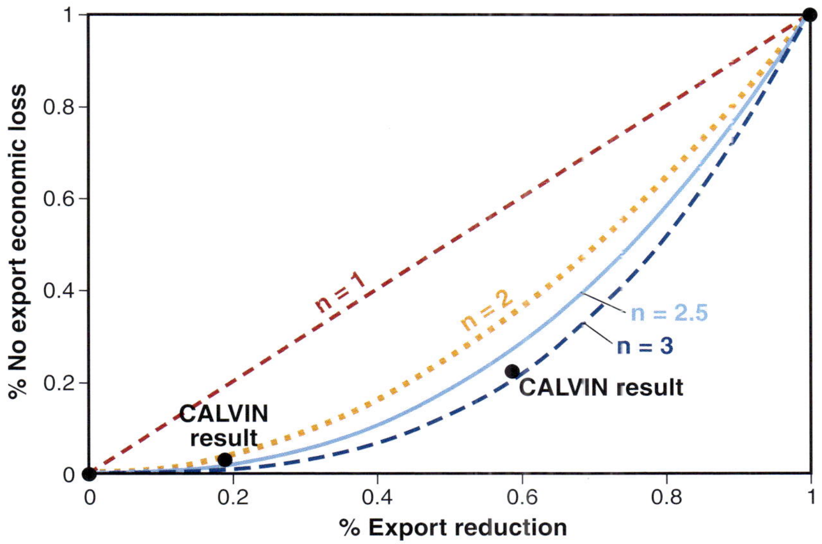

FIGURE A.1 Relationship between proportion of economic loss from reducing exports and proportion of export reduction from recent levels with CALVIN model results.

islands. Here we use one DRMS scenario, which envisions the loss of three major islands with significant damage to four additional islands, to approximate the high-end range of the costs of repairing the five western islands at $2.5 billion.[6]

16. *What exponent relates export reduction to economic cost?*

Figure A.1 shows the relationship between the additional, water-related costs of reducing exports and the proportion of water export quantity reduction for several constants of proportionality with CALVIN model results (Tanaka et al. 2008). From these results, it appears that an exponent between 2 and 3 is reasonable. Any economically efficient allocation of water exports—even at highly reduced levels of exports—will result in an exponent greater than 1. A nonlinear regression of the CALVIN results obtained a value of $n = 2.91$.

ACRONYNMS AND ABBREVIATIONS

BDCP	Bay Delta Conservation Plan
CALFED	state and federal program for the San Francisco Bay and Sacramento–San Joaquin Delta
CALVIN	California Value Integrated Network model
CCWD	Contra Costa Water District
CESA	California Endangered Species Act
cfs	cubic feet per second
Clifton CF	Clifton Court Forebay
CVP	Central Valley Project
CVPIA	Central Valley Project Improvement Act
DRMS	Delta Risk Management Strategy
DWR	Department of Water Resources
EC	electrical conductivity
EPA	U.S. Environmental Protection Agency
ESA	Endangered Species Act (federal)
HCP	Habitat Conservation Plan
maf	million acre-feet

NCCP	Natural Communities Conservation Plan
PC	peripheral canal
PL	Public Law
POD	pelagic organism decline
ppt	parts per thousand
ROD	Record of Decision
SFBCDC	San Francisco Bay Conservation and Development Commission
SWP	State Water Project
SWRCB	State Water Resources Control Board
taf	thousand acre-feet
TRPA	Tahoe Regional Planning Agency
USBR	U.S. Bureau of Reclamation
USGS	U.S. Geological Survey
WAM	water analysis module
WQCP	Water Quality Control Plan

NOTES

CHAPTER 1. INTRODUCTION

1. For the location of individual islands named herein, see map Delta Islands on page xx.

2. See Chapter 4 for details on water use by region.

3. In the first four years of the CALFED investment program, a total of $78 million was spent on levees, only 29 percent of the amount envisioned in the CALFED Record of Decision, signed in August 2000. Total CALFED spending from all sources was $2.5 billion, 66 percent of the level envisaged (Department of Finance, 2005; CALFED, 2000).

CHAPTER 2. THE LEGACIES OF DELTA HISTORY

1. Unless otherwise noted, the discussion in this section draws from Thompson (1957).

2. The Minor Project widened the Sacramento River to 3,500 feet and a mean flood stage of 35 feet. Horseshoe Bend was cut off, Decker Island was created, and a narrow midstream island in front of the city of Rio Vista was removed.

3. Drawing on experience from the 1907 flood, the Major Project proposed 600,000 cubic feet per second (cfs) of discharge capability for the Sacramento River. Creation of the Yolo Bypass was first proposed in a report by Manson and Grunsky for the Public Works Commission in 1894. Other flood control

proposals in this period included that of the Dabney Commission in the early 1900s.

4. The locations of both channels are depicted in Figure 1.2.

5. In 2004, Stockton handled 1.4 percent of total volume and only 0.1 percent of total value of California's sea trade. Sacramento's shares were even lower, at 0.5 percent and 0.06 percent, respectively (www.wisertrade.org)

6. As discussed in Chapter 4, upstream diversions still have major effects on Delta inflows.

7. See Jackson and Paterson (1977) for a discussion of the case.

8. Even in the 1920s, the weakness of Delta levees was seen as a major constraint on Delta solutions, including the design and operation of a saltwater barrier (Young, 1929; Matthew, 1931b).

9. In reaching this conclusion, the Plan's authors drew on several studies conducted in the 1920s, including a 1925 study by the U.S. Bureau of Reclamation (USBR), a 1928 privately financed study on the economics of the barrier (the "Means Report"), a 1929 study for the Department of Public Works (Young 1929), and the report of the joint federal-state commission appointed in 1930 (the Hoover-Young Commission). Among these, the only report to advocate a barrier was the USBR report. See Jackson and Paterson (1977).

10. The proposal was launched in the committee's 1963 report, *Report of the Interagency Delta Committee for Delta Planning* (Jackson and Paterson 1977).

11. In 1978, the SWRCB adopted a new water quality control plan for the Delta and Suisun Marsh (the 1978 Delta Plan) and set new Delta water quality standards with Decision 1485 (D-1485), again focusing on environmental as well as human water quality needs and implying greater restrictions on water exports. Following successful legal challenges at the trial court level, the 1986 "Racanelli Decision" affirmed the SWRCB's broad authority and discretion over water rights and quality issues, including jurisdiction over the CVP. The SWRCB was ordered to prepare a new plan for Delta flows and export guidelines with a greater environmental emphasis. This new draft, put forth in 1988, was withdrawn the following year amid controversy over its legal and water rights implications. Chapter 8 discusses the regulatory issues arising relative to the most recent water quality control plan for the Delta, finalized in 1995 and updated in 2006.

12. In 1970, a preliminary report from the U.S. Geological Survey suggested that the southern San Francisco Bay could suffer from reduced Delta outflows. A 1973 report by the director of the California Department of Fish and Game endorsed the canal for correcting adverse conditions in the Delta for fish (notably problems caused by pumping in the southern Delta), but it also stressed the importance of maintaining adequate flows within the Delta itself and of involving fisheries

agencies in the decision-making process (Arnett 1973). That same year, a student uncovered an unknown, preliminary report from the federal Environmental Protection Agency (EPA) that was highly critical of the canal. The student gave the report to the Friends of the Earth and it was made public. DWR published a 600-page draft *Environmental Impact Report* in August 1974 with only minor changes from the 1969 design. In the early 1970s, environmental groups filed a series of complaints and lawsuits on a range of procedural issues relating to federal involvement and permitting of the peripheral canal (Jackson and Paterson 1977; Hundley 2001).

13. For more on *Paterno*, see Department of Water Resources (2005a).

14. Notably, Senate Bill (SB) 27 in the 2007–08 legislative session and SB 12 in the 2009–10 session, both led by Silicon Valley legislator Joe Simitian.

15. *Natural Resources Defense Council, et al. v. Kempthorne, Findings of Fact and Conclusions of Law Re Interim Remedies Re: Delta Smelt ESA Remand and Reconsultation, United States District Court, Eastern District of California, 1:05-cv-1207 OWW GSA* (2007).

16. Biological opinions are assessments of the consequences of project operations on endangered species. Judge Wanger's rulings invalidated the earlier biological opinion and required a new one to be developed.

17. In *Pacific Coast Federation of Fishermen's Associations, et al. v. Gutierrez, et al.*, 2008 WL 2223070 (E.D. Cal. May 20, 2008) Judge Wanger ruled that the biological opinion for these species was also invalid, and ordered it to be rewritten. This opinion was also issued in December, 2008 (National Marine Fisheries Service, 2008), and decisions have yet to be made regarding required changes in operations.

CHAPTER 3. MANAGING THE INEVITABLE

1. Department of Water Resources (2007a) summarizes expenditures in the Delta Levees Subventions Program and the Delta Levees Special Flood Control Projects program.

2. This analysis excluded two urbanized islands that were included in the analysis for Figure 3.3.

3. In addition to buildings, some infrastructure assets are privately owned (rail lines, electricity and gas lines). Other assets are owned by local public agencies (e.g., the Mokelumne Aqueduct, which belongs to the East Bay Municipal Utility District). Insofar as the values of nonstate assets make it cost-effective to repair levee breaches, it is not clear that state taxpayers (as opposed to landowners, shareholders, and ratepayers) should provide the funds.

4. Although the 2004 Jones Tract repair is often cited as an example of such a case, its total asset value ($550 million) is significantly higher than the costs of repair ($30 million). Current land values are much lower ($42 million). (See the spreadsheet accompanying Suddeth et al. 2008.)

5. This idea underlies recent proposals to create an infrastructure corridor in the southern Delta to consolidate numerous facilities (road, rail, pipelines) and, thereby, economize on levee investments.

6. Unplanned failures of these interior islands during the dry times of the year could still cause water quality problems from a "big gulp"—as saltwater is drawn into the Delta to fill the space in subsided islands. This problem might be largely avoided if the islands either flood during high freshwater inflow periods or are preflooded.

CHAPTER 4. DELTA WATER EXPORTS AND STRATEGIES

1. There is some dispute over water consumption by native vegetation and wetlands under natural conditions. Also, precipitation increases in recent decades might be mitigating some effects of increased water withdrawals (Fox, Mongan, and Miller 1990).

2. Unimpaired flows are estimated using two DWR data series for the period 1956–2005: (1) Dayflow estimates of Delta inflows and exports and (2) estimates of unimpaired or natural Central Valley inflows.

3. Physical seawater barriers have long been considered for the Delta (Department of Water Resources 1930; Matthew 1931a, 1931b) and are common worldwide for dealing with storm surge problems. Lund et al. (2007) rule out permanent barriers because of their likely environmental and economic costs. There has been some recent discussion of closable salinity gates to interrupt the "big gulp" of seawater accompanying major levee failures of subsided islands. This variant would be expensive and unreliable in an earthquake failure. Physical barriers might resolve some water quality problems for water users but would not address worsening ecosystem problems.

4. To date, a few Delta islands have been allowed to remain permanently flooded: Sherman Lake (1875), Franks Tract (1930s), Mildred Island (1983), Liberty Island (1998), and several other smaller islands.

5. Consumptive water uses are diversions of water withdrawn but not returned downstream. Upstream users actually divert a considerably higher volume, but much is returned to the rivers before they reach the Delta.

6. See Department of Water Resources (2008) for cost estimates on through-Delta improvements ranging from $1 billion to $10 billion. A lower estimate of roughly $500 million in improvements was developed by one water agency dependent on Delta exports.

CHAPTER 5. HYDRODYNAMICS AND THE SALINITY OF DELTA WATERS

1. To properly mix salt through the channel network, WAM uses dispersion algorithms developed from more detailed but slower three-dimensional modeling work (Gross et al. 2007).

2. The current EC standards for the Delta are contained in D-1641, adopted in 1999. For an overview, see State Water Resources Control Board (2000) Tables 1, 2 and 4 (also included in Fleenor et al. 2008).

3. For the no-exports simulation, the base case was modified by setting to zero the exports of the CCWD, SWP, CVP, and North Bay Aqueduct (which diverts water at the Barker Slough Pumping Plant).

4. Because these unimpaired flows were estimated using monthly averaged inflow and salinity data, the results are somewhat muted relative to results that would have been obtained using daily data.

5. The western Delta (Chipps Island and Emmaton) would likely have been even more saline in the fall if we had used daily input values.

6. These failures result in little change at Chipps Island and Emmaton, in part since no breaches were included on the Sacramento side of the islands. The location of island breaches affects where water and salts are transported and reintroduced into the Delta with each tidal cycle. Only modest changes occur at Jersey Point, because without the "big gulp" of a sudden levee failure, most saltwater is pulled southward toward the pumps.

7. Since none of the scenarios changes the net Delta outflow, none produces a significant change in salinities at Chipps Island.

8. Although current salinity standards at the Clifton CF location are constant over the irrigation compliance period (at 1,000 S/cm), standards at both Emmaton and Jersey Point vary seasonally and with water year type. Standards are somewhat more stringent at Jersey Point in drier years (State Water Resources Control Board 2000, Table 2).

9. Some studies of statewide water operations and management are presented in Tanaka et al. (2008) and could provide guidance to further hydrodynamic modeling.

CHAPTER 6. WHAT A CHANGING DELTA MEANS FOR THE ECOSYSTEM AND ITS FISH

1. We identify 17 of the Delta's 46 fish species as "desirable" from a management perspective, meaning that they have at least two of the following attributes: They (a) are listed as threatened or endangered, or proposed for listing, under state or federal Endangered Species Acts; (b) support an important sport or commercial fishery; (c) are endemic or native; and (d) depend on the estuary to complete their life cycle, either by living there or migrating through it. The list of desirable species consists mostly of native, especially listed, species (delta smelt, green sturgeon, two runs of Chinook salmon, steelhead). The alien striped bass is also considered desirable because of its fishery and adaptations to estuarine conditions (see Moyle and Bennett 2008 for details).

2. The Delta is just the upper part of the San Francisco Estuary but what occurs in the Delta affects the entire estuary. None of the desirable species is confined just to the Delta; most use other parts of the estuary as well. This chapter focuses on the Delta and Suisun Marsh, but it is important to keep the bigger picture in mind.

3. In the rest of this chapter when we refer to the Delta, we also include Suisun Marsh because its fate and ecosystem are closely tied to those of the Delta proper.

4. This idea is presented in the "Eco-Delta" management alternative described in Lund et al. (2007).

5. This concept is known as "reconciliation ecology" (Rosenzweig 2003).

6. The Delta Wetlands project is a proposal to use two islands in the central Delta as freshwater storage facilities and two others as waterfowl habitat. The model proposes building wide levees that slope toward the interior, supporting riparian vegetation and interior water levels.

7. "Cap-and-trade" refers to a regulatory approach in which a total cap on pollutants is established and polluters are then allowed to trade amongst themselves to determine the actual allocation of the total pollution load. This method can allow pollution reductions to be achieved at a lower overall cost than blanket restrictions on pollutants.

CHAPTER 7. ECONOMICS OF CHANGING WATER SUPPLY AND QUALITY

1. The CALVIN model has been widely applied to provide insights for a variety of California water problems (Jenkins et al. 2003), including climate change with substantial population increases (Tanaka et al. 2006; Medellín-Azuara et al.

2008b), water markets (Jenkins et al. 2004), conjunctive use (Pulido-Velázquez et al. 2004), Hetch Hetchy dam removal (Null and Lund 2006), and Delta policy studies (Tanaka and Lund 2003, Tanaka et al. 2008; Lund et al. 2007).

2. Although pumping costs for conveyance decrease greatly with the end of exports, operating costs from wastewater reuse, desalination, and other pumping increase to a combined higher level.

3. In the absence of water transfers, Berkeley Economic Consulting (2008) estimates that Judge Wanger's December 2007 Interim Remedial Order will result in short-term reductions of about 600,000 acre-feet per year (about 10 percent of pre-decision deliveries), with an estimated average water shortage cost ranging from $140 to $470 million per year, depending on whether water marketing is used to lower the costs of urban shortages. Howitt et al. (2009), examining only the agricultural sector with a different model, finds short-term export cuts to agriculture of 55 to 85 percent, combined with groundwater pumping increases of 25 percent, has an economic income cost to the southern Central Valley of $700 million to $1.55 billion per year, with 16,000 to 37,000 jobs lost, mostly among low wage workers with few alternatives. In the short run, increased groundwater pumping would reduce these severe impacts, and an effective water market could also reduce costs.

4. Confined animal feeding operations (dairies, cattle, etc.) in the Central Valley often use animal wastewater to fertilize feed crops. Salts in wastewater combined with salts in irrigation and groundwater can reduce the allowable rates of wastewater application to the land, which in turn can limit the number of animals that can be supported on a given area of land.

5. These calculations assume that agricultural drainage is reduced in proportion to the imported salt load.

6. This analysis includes the 18 islands in the "do not repair" category in Figure 3.4 (Bacon, Bradford, Coney, Dead Horse, Empire, Holland, Jersey, King, Mandeville, McDonald, Medford, Quimby, Rindge, Staten, Twitchell, Tyler, Venice, and Webb) and one of the indeterminate islands (Wright-Elmwood).

CHAPTER 8. POLICY AND REGULATORY CHALLENGES

1. Hanak and Lund (2008) provides more details on the regulatory issues discussed here.

2. Local levee investments will also be too low if someone else is liable for flood damages. Since the 2003 *Paterno* decision, the state has been held liable for damages in areas behind "project levees" belonging to the Central Valley flood control system, which includes some Delta levees (Department of Water Resources 2005a).

3. Baumol and Willig (1981); Braeutigam (1979); Chessie System Railroads et al. (1981); Damus (1981); and Seneca (1973).

4. Delta cities and counties are members of three separate councils of government—the Association of Bay Area Governments, the Sacramento Council of Governments, and the San Joaquin County Council of Governments.

5. For the second time in its 14-year history, the commission recently ordered a local authority to stop work while it reviews two appeals that challenge development in the primary zone. The case concerns a proposed 162-unit development in the northern Delta town of Clarksburg (unincorporated Yolo County). Litigation could result if there is disagreement between the county and the commission over the project's consistency with the provisions of the Delta Protection Act (Weiser 2006).

6. For information on land trusts and a list of California organizations, see the website of the Land Trust Alliance (www.ltanet.org).

7. See Section 40 CFR.131.10 (h).

8. Case law going back to the early twentieth century has progressively established limits on the extent to which Delta water users are guaranteed water of a certain level of salinity. Salinity standards already vary by water-year type and by location in the Delta in recognition of the excessive costs of meeting higher, uniform standards. It may be possible to modify water quality regulations to allow increasing interannual and seasonal variability by pushing further in this direction—lessening salinity standards in some years (for greater interannual variability) and in later months in the irrigation season (for greater seasonal variability)—without making Delta farming unviable.

9. Arguably, there is strong set of legal tools and precedents to make the case for giving fish and wildlife, especially endangered species, higher priority in setting water quality standards. These tools include the Public Trust Doctrine, Section 5937, of the Fish and Game Code (fish must be in "good condition" below dams), and the 1986 Racanelli Decision (discussed later in this chapter).

10. See the spreadsheet accompanying Suddeth et al. 2008. We have included these costs in our analysis of the costs of not repairing Delta islands after levee failures.

11. There are also questions about whether the state is liable for nonproject levees it does not own, by virtue of subsidies to levee improvements through the Levee Subventions Program (see Chapter 3). Several parties have sued DWR for property damages arising from the 2004 Jones Tract failure on this basis.

12. On the San Joaquin system, some senior users lease water to help moderate flows under the Vernalis Adaptive Management Program. On the Sacramento

system, water users have entered into an agreement to make up to 185,000 acre-feet of water available for in-Valley and export uses through conjunctive use projects. This program is still in the environmental review process.

13. This includes actions by the East Bay Municipal Utility District on the Mokelumne River and actions by the Yuba County Water Agency as part of the recent Yuba River settlement agreement.

14. Data on permit applications are available from http://www.waterboards.ca.gov/ewrims/.

15. In general, HCP requirements are less stringent, so this plan would likely be driven by NCCP requirements.

16. To our knowledge, the only other aquatic HCP is the recently developed multispecies HCP for the lower Colorado River.

17. Despite receiving accolades from the country's planning community, San Diego County's NCCP has been stalled by lawsuits over whether adequate resources are being devoted to its conservation goals.

18. The so-called "God Squad" procedure was established with an amendment to the ESA in 1978. The amendment was prompted by a dam project for the Tennessee Valley Authority, which would have harmed the endangered snail darter (Petersen 2002). In this case, the project did not meet the economic significance justification required to allow an over-ride of the ESA provisions by the God Squad, but Congress granted an exemption. The God Squad exemption was granted for a logging case in the Pacific Northwest, where the species at risk was the northern spotted owl, at the end of the G.H.W. Bush administration. The Clinton administration subsequently determined the timber did not need to be harvested.

19. A third issue, relevant locally and as a motivation for funding the "no" vote campaign, was the costs to southern San Joaquin Valley farming interests of being connected to this new water source (Arax and Wartzman 2003). Judging by the support of valley farmers for a canal this time around, this no longer seems to be an issue, although this support could wane if the canal proposal were too small or too expensive to accommodate farming interests.

20. The proposal was made as part of SB 27 (2007-08 legislative session). Constitutional protections of north coast rivers and Delta water quality were part of the agreement for the peripheral canal proposal in the early 1980s. Dissatisfaction with these environmental protections on the part of some southern Central Valley agricultural interests was a factor in the canal's defeat (Hundley 2001).

21. See Natural Heritage Institute (2008) for a discussion of this issue.

CHAPTER 9. DECISION ANALYSIS FOR DELTA EXPORTS

1. Our analysis finds that a peripheral canal would be built after massive levee failure because this would be the least expensive response. If, instead, the decision was made to rebuild the failed levees or to end exports, the expected cost of the through-Delta strategy would be higher than the range presented here.

2. In Lund et al. (2007) we identified this approach as meriting further investigation. Subsequent hydrodynamic modeling (Fleenor et al. 2008) and improved understanding of the nature of species risks associated with the pumps (Moyle and Bennett 2008) have led us to conclude that this variant on through-Delta pumping is not a viable long-run approach.

APPENDIX

1. *Natural Resources Defense Council et al. v. Kempthorne, Findings of Fact and Conclusions of Law Re Interim Remedies Re: Delta Smelt ESA Remand and Reconsultation.* United States District Court, Eastern District of California, 1:05-cv-1207 OWW GSA (2007).

2. There also will be costs for repairing levees on other Delta islands. We do not include these costs in our estimates here, as they appear unnecessary for water supply. Rather, these costs should be viewed as an element of land and infrastructure policy in the Delta. We estimate annualized repair costs for the subset of islands for which economic activity and infrastructure investments justify these expenditures on the order of $45 million per year (Suddeth et al. 2008).

3. The per-acre-foot cost would increase as the volume of carriage water increases, because higher volumes of cutbacks necessitate reductions in higher valued water uses to meet outflow requirements (Tanaka et al. 2008).

4. Additional carriage water costs to reduce Delta salinity could also be negligible if export restrictions for fish management (Question 8) are already providing these greater outflows.

5. Low estimate: $16.5 million annual levee maintenance plus $150 million annualized costs of western island reinforcement. High estimate: $27.5 million annual levee maintenance plus $250 million annualized costs of western island reinforcement.

6. Estimates of island repair costs for 10 and 30 island failures are $6.3 and $14 billion, respectively (URS and Jack R. Benjamin and Associates 2007b).

GLOSSARY

ADAPTIVE MANAGEMENT A flexible, learning-based management approach in which natural systems are managed to ensure their recovery and improvement, while an understanding of how these systems function is developed to raise the effectiveness of future management actions.

ANADROMOUS FISH SPECIES Fish that live in ocean water and move inland to spawn, such as salmon.

BENEFICIARY PAYS PRINCIPLE The principle that financial responsibility for a project is apportioned among those who benefit from it.

CONSUMPTIVE WATER USE Diversions of water withdrawn but not returned downstream.

CUMULATIVE PROBABILITY The total probability over a range of values or time periods.

DESIRABLE FISH SPECIES Fish with at least two of the following attributes: (a) listed as threatened or endangered, or proposed for listing, under state or federal Endangered Species Acts; (b) support an important sport or commercial fishery; (c) endemic or native; and (d) dependent on the estuary to complete their life cycle, either by living there or migrating through it.

ELECTRICAL CONDUCTIVITY A surrogate measurement for salinity in water.

ENVIRONMENTAL WATER Water allocated to support fish and aquatic habitat, often through minimum flow requirements.

ESTUARY A semi-enclosed embayment where salt water is significantly diluted by fresh water from inflowing rivers.

EXPORT DIVERSIONS Water diverted from the Delta watershed for use in areas to the west and south of the Delta.

EXPORT PUMPS Pumps used for water exports, primarily in the southern Delta.

FISH ENTRAINMENT The drawing of fish or fish larvae into pumps or water diversions.

FISHERY The organized capture of fish for sport or commercial purposes.

GROUND ACCELERATION A measure of intensity of shaking during an earthquake, often described as proportional to the acceleration due to gravity.

GROUNDWATER BANKING The managed storage of water in underground aquifers.

HYDRAULIC FACTORS The effect of water movement on biological and physical processes.

HYDROLOGIC CONDITIONS Conditions related to water inflows.

HYDRODYNAMIC The physics of water movement and the movement of matter (e.g., sediment and salts) in the water.

INDIRECT EXPORTS Water diverted from the Delta watersheds (mainly in the Sacramento Valley and on the east side of the San Joaquin Valley) before the water reaches the Delta. Upstream diversions are a form of indirect exports.

INFLOWS Natural or managed flows of water into a particular location.

INTERTIES OR INTERTIED WATER SYSTEM Connections between water conveyance facilities. An intertied water system such as California's is quite flexible and cost effective in accommodating water shortages or malfunctions at specific water management facilities.

LAND SUBSIDENCE The sinking of lands caused by compaction, oxidation of peat soils, and wind erosion. Many Delta islands have subsided (mostly from oxidation and erosion) to the point at which they now lie many feet below sea level.

MINIMUM FLOW REQUIREMENTS Water flows required by regulators, typically for environmental purposes.

MITIGATION An action intended to moderate some effects of other activities. For instance, flood management agencies often make one-time payments (known as "flood easements") to property owners in areas that will be allowed to flood periodically to help cover the costs of flooding.

OUTFLOW Flows of water going away from a particular location.

PELAGIC FISH SPECIES Fish that live their entire life in open water, above the bottom. Within the Delta, this category includes delta smelt, longfin smelt, and striped bass.

PL 84-99 STANDARDS Minimal standards for levee construction in the Delta to qualify for federal assistance in repairs and rehabilitation.

RECLAMATION The diking and draining of swamplands. Most "reclaimed" Delta lands are used for agriculture, although some lands are used for wildlife habitat and urban development.

SALINITY The concentration of salt in water. As a rough guide, seawater is 35 parts per thousand (ppt)(grams per liter) and fresh water is less than 3.0 ppt. Drinking water is less than 1.0 ppt.

TIDAL EXCURSIONS The mixing of waters caused by daily tidal movements into and out of an estuary.

UNIMPAIRED FLOWS Streamflows unaffected by upstream dams, diversions, or return flows.

UPSTREAM DIVERSIONS Water diverted from the Delta watersheds (mainly in the Sacramento Valley and on the east side of the San Joaquin Valley) before the water reaches the Delta. Upstream diversions are a type of indirect export.

WATER DIVERSIONS The withdrawal of water from a water body, some of which might be returned downstream after use.

WATER EXPORTS Generally refers to water used somewhere other than its area of origin. Direct Delta exports refers to water from Delta watersheds that is sent to points south and west of the Delta. Indirect exports, also known as upstream diversions, refers to water diverted from the Delta watersheds (mainly in the Sacramento Valley and on the east side of the San Joaquin Valley) before it reaches the Delta.

WATER SCARCITY Occurs when water deliveries are less than desired by water users. Scarcity usually implies limited allocations or higher water prices, with associated economic costs and inconvenience for water users. Symptoms of water scarcity might include rationing urban water use, fallowing some farmland, or curtailing recreational activities.

WATER TRANSFERS The exchange, leasing, or permanent sale of the rights to use a particular amount of water from a particular source. Such transfers occur through a "water market," usually involving local and regional water agencies and often state and federal agencies.

WATER YEAR California's water year begins on October 1, the beginning of the rainy season, and ends on September 30.

REFERENCES

Aitchison D. 2007. Institutional options for Delta Vision governance. Letter to John Kirlin, Attorney General's Office, Sacramento, California, dated October 15, 2007. Available from: http://deltavision.ca.gov.

Alexander BS, Mendell GH, Davidson G. 1874. *Report of the Board of Commissioners on the irrigation of the San Joaquin, Tulare, and Sacramento valleys of the State of California.* Washington, D.C.: Government Printing Office.

Anderson K. 2005. *Tending the Wild: Native American Knowledge and the Management of California's Natural Resources,* Berkeley (CA): University of California Press.

Arax M, Wartzman R. 2003. *The king of California: J. G. Boswell and the making of a secret American empire.* New York: Public Affairs.

Arnett GR. 1973. *Maintenance of fish and wildlife in the Sacramento–San Joaquin Estuary in relation to water development.* Sacramento (CA): California Department of Fish and Game.

Atwater BF, Conard SG, Dowden JN, Hedel CW, Donald RL, Savage W. 1979. History, landforms, and vegetation of the estuary's tidal marshes. In: Conomos TJ, Leviton AE, Berson M, editors. *San Francisco Bay: The urbanized estuary.* San Francisco (CA): AAAS, Pacific Division. pp. 347–385.

Baptist MJ, Babovic V, Rodriguez Uthurburu J, Uittenbogaard RE, Mynett A, Verwey A. 2007. On inducing equations for vegetation resistance. *Journal of Hydraulic Research* 45(4): 435–445.

Barnett TD, Pierce DW, Hidalgo HG, Bonfils C, Santer BD, Das T, Bala G, Wood AW, Nozawa T, Mirin AA, Cayan DR, Dettinger MD. 2008. Human-induced changes in the hydrology of the western United States. *Science* 319: 1080–1083.

Baumol WJ, Willig RD. 1981. Fixed costs, sunk costs, entry barriers, and sustainability of monopoly. *Quarterly Journal of Economics* 96(3): 405–431.

Bennett WA. 2005. Critical assessment of the delta smelt population in the San Francisco Estuary, California. San Francisco Estuary and Watershed Science [Internet] 3(2): art1. Available from: http://repositories.cdlib.org/jmie/sfews/vol3/iss2/art1.

Bennett W, Hanak E, Lund J, Moyle P. 2008. An expert survey on the viability of delta fish populations. Appendix E. In: Lund J et al. *Comparing futures for the Sacramento–San Joaquin Delta.* San Francisco (CA): Public Policy Institute of California.

Berkeley Economic Consulting. 2008. *Economic impacts of the Wanger interim order for delta smelt.* Berkeley (CA): Berkeley Economic Consulting. 40 p.

Berkes F, Folke C. 1998. *Linking social and ecological systems: Management practices and social mechanisms for building resilience.* Cambridge, UK: Cambridge University Press.

Bindoff NL, Willebrand J, Artale V, Cazenave A, Gregory J, Gulev S, Hanawa K, Le Quéré C, Levitus S, Nojiri Y, Shum CK, Talley LD, Unnikrishnan A. 2007. Observations: Oceanic climate change and sea level. In: Solomon S, Qin D, Manning M, Chen Z, Marquis M, Averyt KB, Tignor M, Miller HL, editors. *Climate change 2007: The physical science basis.* Contribution of working group I to the fourth assessment report of the Intergovernmental Panel on Climate Change. Cambridge, UK: Cambridge University Press.

Braeutigam RR. 1979. Optimal pricing with intermodal competition. *American Economic Review* 69(1): 38–49.

Burau, JR. "DRERIP Hydrodynamic Transport Conceptual Model," draft report, 12/20/2007.

CALFED. 2000. *California's water future: A framework for action.* Sacramento (CA): CALFED Bay-Delta Program.

California Secretary of State. 1982. Statement of vote, primary election, June 8, 1982. Sacramento (CA): California Secretary of State.

Cayan DR, Bromirski PD, Hayhoe K, Tyree M, Dettinger MD, Flick RE. 2008a. Climate change projections of sea level extremes along the California coast. *Climatic Change* 87 (Suppl. 1): S57–S73.

Cayan DR, Maurer EP, Dettinger MD, Tyree M, Hayoe K. 2008b. Climate change scenarios for the California region. *Climatic Change* 87 (Suppl. 1): S21–S42.

Chen W-H, Haunschild K, Lund J. 2008. Delta drinking water quality and treatment costs. Appendix H. In: Lund J et al. *Comparing futures for the Sacramento–San Joaquin Delta.* San Francisco (CA): Public Policy Institute of California.

Chessie System Railroads, Consolidated Rail Corporation, Family Lines Rail System, Norfolk and Western Railway Company, Southern Railway System. 1981. Comments of five railroads. *Ex Parte* 347 (Sub-No. D). Vol. I. p. 111–112.

Church JA, White NJ. 2006. Twentieth century acceleration in sea level rise. *Geophysical Research Letters* 33: LO1602.

Cohon JL. 1978. *Multiobjective programming and planning.* New York: Academic Press.

Contra Costa Water District. 2005. *Urban water management plan.* Concord (CA): Contra Costa Water District.

Damus S. 1981. Two-part tariffs and optimum taxation: The case of railway rates. *American Economic Review* 71(1): 65–79.

Davis MB, Shaw RG. 2001. Range shifts and adaptive responses to quaternary climate change. *Science* 292: 673–679.

de Alth S, Rueben K. 2005. *Understanding infrastructure financing for California. Occasional paper.* San Francisco (CA): Public Policy Institute of California.

de Neufville R. 1990. Applied systems analysis. New York: McGraw-Hill.

Department of Boating and Waterways. 2002. *Sacramento–San Joaquin Delta boating needs assessment.* Sacramento (CA): Dept. of Boating and Waterways.

Department of Finance. 2005. *A Fiscal Review: CALFED Bay-Delta Program summary of expenditures as of September 30, 2004.* Sacramento (CA): Office of State Audits and Evaluations.

Department of Public Works. 1923. Water resources of California. A report to the Legislature of 1923. Public Works Bulletin No. 4. Sacramento (CA): Dept. of Public Works.

Department of Public Works. 1930. The state water plan. A report to the Legislature of 1931. Public Works Bulletin No. 25. Sacramento (CA): Dept. of Public Works, Division of Water Resources.

Department of Water Resources. 1957. California water plan. Bulletin No. 3. Sacramento (CA): Dept. of Water Resources.

Department of Water Resources. 1995. *Sacramento–San Joaquin Delta water atlas.* Sacramento (CA): Dept. of Water Resources.

Department of Water Resources. 1998. California water plan update. Bulletin 160-98. Sacramento (CA): Dept. of Water Resources.

Department of Water Resources. 2005a. *Flood warnings: Responding to California's flood crisis.* Sacramento (CA): Dept. of Water Resources.

Department of Water Resources. 2005b. California water plan update Bulletin 160-05. Sacramento (CA): Dept. of Water Resources.

Department of Water Resources. 2007a. Delta Levees Program components [Internet]. Available from: http://www.dfm.water.ca.gov/dsmo/levees/components.html.

Department of Water Resources. 2007b. DWR releases water delivery impact estimates following Wanger decision. Advisory. Sacramento (CA): Dept. of Water Resources.

Department of Water Resources. 2007c. Delta Risk Management Strategy Program [Internet]. Available from: http://www.drms.water.ca.gov/.

Department of Water Resources. 2008. An initial assessment of dual Delta water conveyance. Prepared for the Delta Vision Blue Ribbon Task Force. Sacramento (CA): Dept. of Water Resources.

Drexler JZ, de Fontaine CS, Knifong DL. 2007. Age determination of the remaining peat in the Sacramento–San Joaquin Delta, California, USA. USGS Open File Report 2007-1303. Sacramento (CA): U.S. Geological Survey.

East Bay Municipal Utility District. 2005. Urban water management plan. Oakland (CA): East Bay Municipal Utility District.

Eisenstein W, Kondolf M, Cain J. 2007. Re-envisioning the Delta: Alternative futures for the heart of California. IURD report # WP-2007-01. Berkeley (CA): Dept. of Landscape Architecture and Environmental Planning, University of California, Berkeley.

Feyrer F, Nobriga, ML, Sommer, TR. 2007. Multi-decadal trends for three declining fish species: Habitat patterns and mechanisms in the San Francisco Estuary, California, USA. *Canadian Journal of Fisheries and Aquatic Sciences* 64(4): 723–734.

Fleenor W, Hanak E, Lund J, Mount J. 2008. Delta hydrodynamics and water quality with future conditions. Appendix C. In: Lund J et al. *Comparing futures for the Sacramento–San Joaquin Delta.* San Francisco (CA): Public Policy Institute of California.

Florsheim JL, Dettinger MD. 2007. Climate and floods still govern California levee breaks. *Geophysical Research Letters* 34: L22403.

Folke C, Carpenter S, Walker B, Scheffer M, Elmquist T, Gunderson L, Holling CS. 2004. Regime shifts, resilience, and biodiversity in ecosystem management. *Annual Reviews in Ecology and Systematics* 35: 557–581.

Fox JP, Mongan TR, Miller WJ. 1990. Trends in freshwater inflow to San Francisco Bay from the Sacramento–San Joaquin Delta. *Water Resources Bulletin* 26(1): 101–116.

Gross ES, MacWilliams ML, Nidzieko N. 2007. Three-dimensional salinity simulations of sea level rise scenarios. In: *Department of Water Resources. Delta Risk Management Strategy.* Sacramento (CA): Department of Water Resources.

Grunwald M. 2006. *The swamp: The Everglades, Florida, and the politics of paradise.* New York: Simon and Schuster.

Gunderson LH, Holling CS, editors. 2001. *Panarchy: Understanding transformation in human and natural systems.* Washington, D.C.: Island Press.

Hanak E. 2003. *Who should be allowed to sell water in California? Third party issues and the water market.* San Francisco (CA): Public Policy Institute of California.

Hanak E, Lund J. 2008. Policy and regulatory challenges for the Delta. Appendix A. In: Lund J et al. *Comparing futures for the Sacramento–San Joaquin Delta.* San Francisco (CA): Public Policy Institute of California.

Herbold B, Moyle PB. 1989. Ecology of the Sacramento–San Joaquin Delta: A community profile. U.S. Fish and Wildlife Service Biological Report, Vol. 85. No. 7.22.

Hobbs BF, Chao PT, Venkatesh BN. 2004. Using decision analysis to include climate change in water resources decision making. *Climatic Change* 37: 177–202.

Howitt RE, MacEwan D, Medellin-Azuara J, Lund J. 2008. Economic impacts of reductions in Delta exports on Central Valley agriculture. Technical memorandum. Davis (CA): Dept. of Agricultural and Resource Economics, University of California, Davis. 20 p.

Howitt RE, Medellin-Azuara J, MacEwan DJ. 2009. Measuring the employment impacts of water reductions. Technical Note Department of Agriculture and Resource Economics and Center for Watershed Sciences. University of California, Davis CA.

Hundley N, Jr. 2001. *The great thirst. Californians and water: A history.* Revised edition. Berkeley (CA): University of California Press.

Illingworth W, Mann R, Hatchet S. 2005. Economic consequences of water supply export disruption due to seismically initiated levee failures in the Delta. Appendix B. In: Jack R. Benjamin & Associates. *Preliminary seismic risk analysis associated with levee failures in the Sacramento–San Joaquin Delta.* Menlo Park (CA): Jack R. Benjamin & Associates.

Intergovernmental Panel on Climate Change. 2007. *Climate change 2007: Synthesis report. Contribution of Working Groups I, II and III to the Fourth Assessment.* Pachauri RK, Reisinger A, editors. Geneva, Switzerland: IPCC. 104 p.

Isenberg P, Florian M, Frank RM, McKernan T, McPeak SW, Reilly WK, Seed R. 2008a. *Blue Ribbon Task Force Delta vision: Our vision for the California Delta.* Sacramento (CA): State of California Resources Agency.

Isenberg P, Florian M, Frank RM, McKernan T, McPeak SW, Reilly WK, Seed R. 2008b. *Delta Vision Strategic Plan.* Sacramento (CA): Delta Vision Blue Ribbon Task Force.

Jackson WT, Paterson AM. 1977. The Sacramento–San Joaquin Delta and the evolution and implementation of water policy: An historical perspective. California Water Resources Center, Contribution No. 163. Davis (CA): University of California, Davis.

Jassby AD, Cloern JE. 2000. Organic matter sources and rehabilitation of the Sacramento–San Joaquin Delta (California, USA). *Aquatic Conservation: Marine and Freshwater Ecosystems* 10: 323–352.

Jenkins MW, Draper AJ, Lund JR, Howitt RE, Tanaka SK, Ritzema R, Marques GF, Msangi SM, Newlin BD, Van Lienden BJ, Davis MD, Ward KB. 2001. Improving California water management: Optimizing value and flexibility. Center for Environmental and Water Resources Engineering Report No. 01-1. Davis (CA): Department of Civil and Environmental Engineering, University of California, Davis. Available from: http://cee.engr.ucdavis.edu/faculty/lund/CALVIN/.

Jenkins MW, Lund JR, Howitt RE, Draper AJ, Msangi SM, Tanaka SK, Ritzema RS, Marques GF. 2004. Optimization of California's water system: Results and insights. *Journal of Water Resources Planning and Management* 130(4): 271–280.

Kelley R. 1989. *Battling the inland sea.* Berkeley (CA): University of California Press.

Kimmerer WL. 2008. Losses of Sacramento River Chinook salmon and delta smelt to entrainment in water diversions in the Sacramento–San Joaquin Delta. San Francisco Estuary and Watershed Science [Internet]. 6(2): art2. Available from: http://repositories.cdlib.org/jmie/sfews/vol6/iss2/art2

Knowles N. 2002. Natural and management influences on freshwater inflows and salinity in the San Francisco Estuary at monthly to interannual scales. *Water Resources Research* 38(12): 1–11.

Krone RB. 1979. Sedimentation in the San Francisco Bay system. In: Conomos TJ, Leviton AE, Berson M, editors. *San Francisco Bay: The urbanized estuary.* San Francisco (CA): AAAS, Pacific Division. pp. 85–96.

Little Hoover Commission. 2005. *Still imperiled, still important. The Little Hoover Commission's review of the CALFED Bay-Delta Program.* Sacramento (CA): Little Hoover Commission.

Liu J, Thomas D, Carpenter S, Folke C, Alberti M, Redman C, Schneider S, Ostrom E, Pell A, Lubchenco J, Taylor W, Ouyang A, Deadman P, Kratz T, Provencher W. 2007. Coupled human and natural systems. *Ambio* 36(8): 639–649.

Logan SH. 1989. An economic analysis of flood control policy in the Sacramento–San Joaquin Delta, Contribution 199. Davis (CA): California Water Resources Center.

Logan SH. 1990a. Global warming and the Sacramento–San Joaquin Delta, California. *Agriculture* 44(3): 16–18.

Logan SH. 1990b. Simulating costs of flooding under alternative policies for the Sacramento–San Joaquin River Delta. *Water Resources Research* 26(5): 799–809.

Ludwig D, Hilborn R, Walters C. 1993. Uncertainty, resource exploitation, and conservation: lessons from history. *Science* 260(5104): 17–36.

Lund JR. 2008. Peripheral canal design and implementation options. Appendix G. In: Lund J et al. *Comparing futures for the Sacramento–San Joaquin Delta*. San Francisco (CA): Public Policy Institute of California.

Lund JR. 2009. Probabilistic design and optimization class notes. ECI 249. Davis (CA): Dept. of Civil and Environmental Engineering, University of California, Davis. Available from: http://cee.engr.ucdavis.edu/faculty/lund/Classes/ECI249Notes.pdf

Lund J, Hanak E, Fleenor W, Howitt R, Mount J, Moyle P. 2007. *Envisioning futures for the Sacramento–San Joaquin Delta*. San Francisco (CA): Public Policy Institute of California.

Lund J, Hanak E, Fleenor W, Bennett W, Howitt R, Mount J, Moyle P. 2008a. *Comparing futures for the Sacramento–San Joaquin Delta*. San Francisco (CA): Public Policy Institute of California.

Lund J, Hanak E, Fleenor W, Bennett W, Howitt R, Mount J, Moyle P. 2008b. Decision analysis of Delta strategies. Appendix J. In: Lund J et al. *Comparing futures for the Sacramento–San Joaquin Delta*. San Francisco (CA): Public Policy Institute of California.

Maguire LA. 2004. What can decision analysis do for invasive species management? *Risk Analysis* 24(4): 859–68.

Malamud-Roam F, Dettinger M, Lynn Ingram B, Hughes MK, Florsheim JL. 2007. Holocene climates and connections between the San Francisco Bay Estuary and its watershed: A review. San Francisco Estuary and Watershed Science [Internet]. 5(1): art3. Available from: http://repositories.cdlib.org/jmie/sfews/vol5/iss1/art3.

Malcolm JR, Liu C, Neilson RP, Hansen L, Hannah L. 2006. Global warming and extinctions of endemic species from biodiversity hotspots. *Conservation Biology* 20(2): 538–548.

Matthew R. 1931a. Variation and control of salinity in Sacramento–San Joaquin Delta and upper San Francisco Bay. Bulletin 27. Sacramento (CA): Division of Water Resources, California Department of Public Works.

Matthew R. 1931b. Economic aspects of a salt water barrier below the confluence of Sacramento and San Joaquin rivers. Bulletin 28. Sacramento (CA): Division of Water Resources, California Department of Public Works.

Maurer EP. 2007. Uncertainty in hydrologic impacts of climate change in the Sierra Nevada Mountains, California under two emissions scenarios. *Climatic Change* 82: 309–325.

Medellin-Azuara J, Harou JJ, Olivares MA, Madani-Larijani K, Lund JR, Howitt RE, Tanaka S. K, Jenkins MW, Zhu T. 2008a. Adaptability and adaptations of California's water supply system to dry climate warming. *Climatic Change* 87(Suppl.1): S75–S90.

Medellin-Azuara J, Howitt R, Lund J, Hanak, E. 2008b. Economic effects on agriculture of water export salinity south of the Sacramento–San Joaquin Delta. Appendix I. In: Lund J et al. *Comparing futures for the Sacramento–San Joaquin Delta.* San Francisco (CA): Public Policy Institute of California.

Meehl GA, Stocker TF, Collins WD, Friedlingstein P, Gaye AT, Gregory JM, Kitoh A, Knutti R, Murphy JM, Noda A, Raper SCB, Watterson IG, Weaver AJ, Zhao Z-C. 2007. Global climate projections. In: Solomon S, Qin D, Manning M, Chen Z, Marquis M, Averyt KB, Tignor M, Miller HL, editors. Climate change 2007: The physical science basis. Contribution of Working Group I to the fourth assessment report of the Intergovernmental Panel on Climate Change.

Misczynski D. 2009. *Fixing the Delta: How Will We Pay for It?* San Francisco (CA): Public Policy Institute of California.

Mount J. 2007. Sea level rise and Delta planning. Memo from the CALFED Independent Science Board to Mike Healey, CALFED Lead Scientist, dated September 6, 2007. Available from: http://calwater.ca.gov/science/pdf/isb/meeting_082807/ISB_response_to_ls_sea_level_090707.pdf.

Mount JF, Twiss R. 2005. Subsidence, Sea Level Rise, Seismicity in the Sacramento–San Joaquin Delta. San Francisco Estuary and Watershed Science [Internet]. 3(1): art5. Available from: http://repositories.cdlib.org/jmie/sfews/vol3/iss1/art5.

Moyle PB. 2002. *Inland fishes of California, revised and expanded.* Berkeley (CA): University of California Press.

Moyle PB. 2008. The future of fish in response to large-scale change in the San Francisco Estuary, California. In: McLaughlin KD, editor. *Mitigating impacts of natural hazards on fishery ecosystems.* American Fishery Society, Symposium 64. Bethesda (MD): American Fishery Society.

Moyle PB, Bennett WA. 2008. Future of the Delta ecosystem and its fishes. Appendix D. In: Lund J et al. *Comparing futures for the Sacramento–San Joaquin Delta.* San Francisco (CA): Public Policy Institute of California.

National Marine Fisheries Service. 2008. *Draft biological opinion on the long-term Central Valley Project and State Water Project operations criteria and plan.* Long Beach (CA): National Marine Fisheries Service, Southwest Regional Office. Available from: http://swr.nmfs.noaa.gov/sac/myweb8/BiOpFiles/2009/Draft_OCAP_Opinion.pdf

Natural Heritage Institute. 2008. *Concept offered for consideration by BDCP Implementation Structure/Governance Work Group.* San Francisco (CA): Natural Heritage Institute.

Natural Resources Agency. 2009. *Delta Vision Committee implementation report.* Sacramento (CA): California Natural Resources Agency.

Null S, Lund JR. 2006. 2006. Re-assembling Hetch Hetchy: Water supply implications of removing O'Shaughnessy Dam. *Journal of the American Water Resources Association* 42(4): 395–408.

Orlob GT. 1982. An alternative to the peripheral canal. *Journal of the Water Resources Planning and Management Division* 108(WR1): 123–141.

Orlob GT. 1991. San Joaquin salt balance: Future prospects and possible solutions. In: Dinar A, Zilberman D, editors. *The economics and management of water and drainage in agriculture.* Boston (MA): Kluwer. p. 143–167.

Owen D. 2007. Law, environmental dynamism, reliability: The rise and fall of CALFED. *Environmental Law* 77: 1145–1215.

Parma AM, National Center for Ecological Analysis and Synthesis Working Group on Population Management. 1998. What can adaptive management do for our fish, forests, food, and biodiversity? *Integrative Biology* 1: 16–26.

Petersen S. 2002. *Acting for endangered species: The statutory ark.* Lawrence (KS): University Press of Kansas.

Plater J, Wade WW. 2002. Estimating potential demand for freshwater recreation activities in the Sacramento–San Joaquin Delta 1997–2020. Columbia (TN): Energy and Water Economics. Available from: http://www.dbw.ca.gov.

Port of Sacramento. 2006. Overview. Sacramento (CA): Port of Sacramento. Available from: http://www.portofsacramento.com.

Pulido-Velázquez M, Jenkins MW, Lund JR. 2004. Economic values for conjunctive use and water banking in Southern California. *Water Resources Research* 40(3): 1–15.

Rahmstorf S. 2007. A semi-empirical approach to projecting future sea-level rise. *Science* 315: 368–370.

Ramsey F. 1927. A contribution to the theory of taxation. *Economic Journal* 37: 47–61.

ReVelle CS, Whitlach EE, Wright JR. 1997. *Civil and environmental systems engineering.* Upper Saddle River (NJ): Prentice-Hall.

Rosenzweig M. 2003. *Win-win ecology: How the Earth's species can survive in the midst of human enterprise.* Oxford (UK): Oxford University Press.

San Francisco Public Utilities Commission. 2005. Urban water management plan. San Francisco (CA): San Francisco Public Utilities Commission. Available from: http://www.sfwater.org.

Santa Clara Valley Water District. 2005. Urban water management plan. San Jose (CA): Santa Clara Valley Water District. Available from: http://www.valleywater.org.

Sax JL. 2002. Review of the laws establishing the SWRCB's permitting authority over appropriations of groundwater classified as subterranean streams and the SWRCB's implementation of those laws. SWRCB No. 0-076-300-0. Sacramento (CA): State Water Resources Control Board.

Scheffer M, Carpenter SR. 2003. Catastrophic regime shifts in ecosystems: Linking theory to observation. *Trends in Ecology and Evolution* 18: 648–656.

Scheffer M, Carpenter S, Foley JA, Folkes C, Walker B. 2001. Catastrophic shifts in ecosystems. *Nature* 413: 591–596.

Schoups G. 2004. Regional-scale hydrologic modeling of subsurface water flow and reactive salt transport in the western San Joaquin Valley, California [PhD dissertation]. Available from: University of California, Davis.

Seneca RS. 1973. Inherent advantage, costs and resource allocation in the transportation industry. *American Economic Review* 63(5): 945–956.

Shlemon RJ, Begg EL. 1975. Late Quaternary evolution of the Sacramento–San Joaquin Delta, California. In: Suggate RP, Cressel MM, editors. Quaternary studies. Bulletin 13. The Royal Society of New Zealand. pp. 259–266.

Snow L. 2006. Protecting Sacramento/San Joaquin Bay–Delta water supplies and responding to failures in California water deliveries. Testimony before the U.S. House of Representatives Committee on Resources, Subcommittee on Water and Power. Sacramento (CA): Dept. of Water Resources. Available from: http://www.publicaffairs.water.ca.gov.

State Water Resources Control Board. 1999. *Final Environmental Impact Report for Implementation of the 1995 Bay/Delta Water Quality Control Plan*. Sacramento (CA): State Water Resources Control Board, Bay Delta Program.

State Water Resources Control Board. 2000. Revised Water Right Decision D-1641. Sacramento (CA): California Environmental Protection Agency. Available from: http://www.waterrights.ca.gov/Decisions/D1641revs.pdf.

Stewart I, Cayan DR, Dettinger MD. 2004. Changes in snowmelt runoff timing in western North America under a "business as usual" climate change scenario. *Climatic Change* 62: 217–232.

Suddeth R. 2009. Levee decisions and sustainability for the Sacramento San Joaquin Delta. [Masters Thesis]. University of California, Davis.

Suddeth R, Mount J, Lund J. 2008. Levee decisions and sustainability for the Sacramento–San Joaquin Delta. Appendix B. In: Lund J et al. *Comparing futures for the Sacramento–San Joaquin Delta*. San Francisco (CA): Public Policy Institute of California.

Tanaka SK, Lund JR. 2003. Effects of increased Delta exports on Sacramento Valley's economy and water management. *Journal of the American Water Resources Association* 39(6): 1509–1519.

Tanaka SK, Zhu T, Lund JR, Howitt RE, Jenkins MW, Pulido MA, Tauber M, Ritzema RS, Ferreira IC. 2006. Climate warming and water management adaptation for California. *Climatic Change* 76(3–4).

Tanaka S, Connell C, Madani K, Lund J, Hanak E. 2008. Economic costs and adaptations for increasing Delta outflows and reducing or ending Delta exports. Appendix F. In: Lund J et al. *Comparing futures for the Sacramento–San Joaquin Delta*. San Francisco (CA): Public Policy Institute of California.

Thompson J. 1957. Settlement geography of the Sacramento–San Joaquin Delta, California [PhD dissertation]. Available from: Stanford University.

Torres RA, et al. 2000. Seismic vulnerability of the Sacramento–San Joaquin Delta Levees. Report of the Levees and Channels Technical Team, Seismic Vulnerability Sub-Team, to the CALFED Bay-Delta Program. Sacramento(CA): CALFED Bay-Delta Program.

URS Corporation, Jack R. Benjamin and Associates, Inc. 2007a. Technical memorandum, topical area: Water analysis module. Prepared for the California Department of Water Resources.

URS Corporation, Jack R. Benjamin and Associates, Inc. 2007b. Delta Risk Management Strategy (DRMS) phase 1 risk analysis report, draft. Prepared for the California Department of Water Resources.

URS Corporation ,J.R. Benjamin and Associates, Inc. 2009. Delta Risk Management Strategy (DRMS) Risk Analysis Report. Prepared for the California Department of Water Resources.

U.S. Bureau of Reclamation. 2005. CALSIM II San Joaquin River model [draft—online version]. Sacramento (CA): U.S. Department of the Interior, Mid-Pacific Region. Available from: http://science.calwater.ca.gov/pdf/calsim/CALSIMSJR_DRAFT_072205.pdf.

U.S. Fish and Wildlife Service. 2008. Biological opinion on the long-term operational criteria and plan (OCAP) for coordination of the Central Valley Project and State Water Project. Sacramento (CA): U.S. Fish and Wildlife Service.

U.S. Geological Survey. 2003. Earthquake probabilities in the San Francisco Bay region, 2002–2031. Working Group on California Earthquake Probabilities. Open-File Report 03-214.

van Dantzig D. 1956. Economic decision problems for flood prevention. *Econometrica* 24(3): 276–287.

Voortman HG, van Gelder PHAJM, Vrijling JK. 2002. Risk-based design of large-scale flood defense systems. 28th International Conference on Coastal Engineering; Cardiff; July 2002. pp. 2373–2385.

Walker B, Salt D. 2006. *Resilience thinking: Sustaining ecosystems and people in a changing world.* Washington, D.C.: Island Press.

Weiser M. 2006a. Sugar mill project halted; two groups appeal Yolo's approval of Clarksburg Tract, citing Delta risks. *Sacramento Bee,* November 8, 2006.

Whitney V. 2008. Presentation of the Deputy Director for Water Rights, State Water Resources Control Board, to the Delta Vision Blue Ribbon Task Force, January 31, 2008. Available from: http://deltavision.ca.gov/.

Young WR. 1929. Report on salt water barrier below confluence of Sacramento and San Joaquin rivers, California. Bulletin 22, Vols. I and II. Sacramento (CA): California Department of Public Works, Division of Water Resources.

INDEX

Page numbers in **bold type** refer to maps, charts, and tables. Page numbers followed by *n* refer to endnotes.

FRESHWATER ECOLOGY SERIES VOLUMES

WWW.UCPRESS.EDU/GO/FWE

Freshwater Mussel Ecology: A Multifactor Approach to Distribution and Abundance
David L. Strayer

Mirror Lake: Interactions among Air, Land, and Water
Thomas C. Winter and Gene E. Likens, eds.

Comparing Futures for the Sacramento–San Joaquin Delta
Jay R. Lund, Ellen Hanak, William E. Fleenor, William A. Bennett, Richard E. Howitt, Jeffrey F. Mount, and Peter B. Moyle

Indexer: Indexing Solutions
Composition: Michael Bass Associates
Text and display: Bembo
Printer and binder: Sheridan Books, Inc.